ENERGY REVOLUTION

About Island Press

Island Press is the only nonprofit organization in the United States whose principal purpose is the publication of books on environmental issues and natural resource management. We provide solutions-oriented information to professionals, public officials, business and community leaders, and concerned citizens who are shaping responses to environmental problems.

In 2003, Island Press celebrates its nineteenth anniversary as the leading provider of timely and practical books that take a multidisciplinary approach to critical environmental concerns. Our growing list of titles reflects our commitment to bringing the best of an expanding body of literature to the environmental community throughout North America and the world.

Support for Island Press is provided by The Nathan Cummings Foundation, Geraldine R. Dodge Foundation, Doris Duke Charitable Foundation, Educational Foundation of America, The Charles Engelhard Foundation, The Ford Foundation, The George Gund Foundation, The Vira I. Heinz Endowment, The William and Flora Hewlett Foundation, Henry Luce Foundation, The John D. and Catherine T. MacArthur Foundation, The Andrew W. Mellon Foundation, The Moriah Fund, The Curtis and Edith Munson Foundation, National Fish and Wildlife Foundation, The New-Land Foundation, Oak Foundation, The Overbrook Foundation, The David and Lucile Packard Foundation, The Pew Charitable Trusts, The Rockefeller Foundation, The Winslow Foundation, and other generous donors.

The opinions expressed in this book are
those of the author(s) and do not necessarily
reflect the views of these foundations.

ENERGY REVOLUTION

POLICIES FOR A SUSTAINABLE FUTURE

HOWARD GELLER

ISLAND PRESS
Washington • Covelo • London

Library of Congress Cataloging-in-Publication Data.

Geller, Howard S.
Energy revolution : policies for a sustainable future / Howard Geller.
p. cm.
Includes bibliographical references and index.
ISBN 1-55963-964-4 (cloth : alk. paper) — ISBN 1-55963-965-2 (pbk. : alk. paper)
1. Energy policy. 2. Renewable energy sources. 3. Sustainable development.
4. Energy policy—United States. 5. Energy policy—Brazil. I. Title.
HD9502.A2 G42 2003
333.79—dc21 2002015749

British Cataloguing-in-Publication Data available.

Printed on acid-free, recycled paper

Manufactured in the United States of America
09 08 07 06 05 04 03 8 7 6 5 4 3 2 1

I dedicate this book to my teachers and mentors, starting with my parents, Norma and Albert Geller, who taught me to strive for excellence in whatever I pursue; to my professors at Clark University, especially Rob Goble, Chris Hohenemser, Roger Kasperson, and Mike McClintock, who sparked my interest in energy issues; to my professors at Princeton University, especially Hal Feiveson, Rob Socolow, Ted Taylor, Frank von Hippel, and Bob Williams, who taught me to conduct rigorous analysis and challenge "conventional wisdom"; to José Goldemberg and Amulya Reddy, who opened my heart and mind to the problems of developing nations; to Carl Blumstein, Henry Kelly, and Marc Ross, who guided me as my career as an energy analyst took shape; and to Art Rosenfeld, who has been an inspiration throughout. I have been fortunate to have had the opportunity to learn from all of you.

Contents

Preface *ix*

Acknowledgments *xiii*

1 Introduction 1

2 Barriers 33

3 Policy Options 47

4 Market Transformation 93

5 The United States: Policies and Scenarios 131

6 Brazil: Policies and Scenarios 165

7 International Policies and Institutions 191

8 Toward a Sustainable Energy Future 215

Appendix *243*

References *245*

Index *277*

Preface

I began writing this book in 1998 when energy prices were relatively low and energy was not a major issue. The United States and most other wealthy nations were experiencing robust economic growth. Policy makers were grappling with the economic crisis in Asia. Global warming, however, was one issue garnering international attention. With emissions of the gases causing global warming on the rise worldwide, nations joined together to limit future emissions through the Kyoto Protocol. But it was far from clear if the emissions targets contained in the Protocol were achievable, and if so what the path to first meeting and then exceeding these emissions targets might resemble.

Much has transpired to raise the profile of energy issues since 1998. Oil prices climbed, California and nearby states experienced electricity price shocks and shortages, and the collapse of Enron devastated financial markets. In the United States, the Bush administration, U.S. Congress, and various interest groups have been debating major energy legislation. And terrorism sponsored by the al-Qaeda network has reminded us of our heavy dependence on energy imports, our vulnerability to oil supply disruptions or price shocks, and the possibility of terrorist attacks on key components of our energy infrastructure.

This book comes at a critical juncture. Evidence of human-induced global warming keeps mounting. Many nations, regions, and companies have made significant progress in reducing their emissions of the so-called greenhouse gases causing this problem. Markets for cleaner energy technologies such as wind turbines, solar photovoltaic systems, and compact fluorescent lamps are booming. On the other hand, there are powerful forces pushing to maintain the status quo—continued heavy reliance on fossil fuels including coal and petroleum, and consequently rising greenhouse gas emissions and intensifying global warming. Which direction will the world take?

Various other books and studies have suggested that a transition away from fossil fuels to renewable energy sources coupled with greater energy efficiency—a clean energy revolution—is desirable and feasible. But most concentrate on the technologies for achieving this transition, not the policies and strategies. I believe, however, the real challenge we face is not technological. Most of the technologies for a clean energy revolution are either commercially available or emerging in the marketplace. The bigger challenge is how to overcome the barriers preventing widespread adoption of these technologies in the coming decades; how to implement a clean energy revolution.

This book is intended to fill this gap. There has been a tremendous amount of experience with policies for advancing energy efficiency and renewable energy technologies over the past decade or two. It is now possible to envision how an energy revolution could be accomplished during this century. This book reviews the policy experience, including the lessons from past efforts to advance energy efficiency and renewable energy use. It also recommends and analyzes policies that could be adopted to facilitate a clean energy revolution in the future.

This book attempts to be holistic from a number of perspectives. First, it covers energy policy experience and options in both industrialized and developing nations. In fact I have chosen to interweave the geographic coverage since leadership on energy policy and thus important lessons have been provided by rich, middle-income, and poor nations. Second, it examines the range of economic, social, and environmental threats posed by a "business-as-usual" energy future, and likewise the range of benefits offered by a clean energy revolution. Global warming is considered, but so are local and regional air pollution, security risks, Third World poverty, and global equity. And third, it covers policies for implementing greater energy efficiency and renewable energy sources. Both strategies are critical for achieving a clean energy revolution.

I hope this book will appeal to a broad audience that includes policy makers, business leaders, environmentalists, and students. I also hope it will appeal to readers in many nations, not just the United States. This book is about moving toward a sustainable energy future, something all citizens of the world have a stake in and something that all nations must work cooperatively to achieve. Recognizing that the scope is very broad, I provide many references for those seeking more information on particular policies or issues.

A Word on Energy Units

I have chosen to use the conventional energy units from a particular country or region (e.g., quadrillion Btus (quads) in the United States, million tons of coal equivalent in China, and million tons of oil equivalent in Brazil). These units are familiar to the readers in each region. The following section presents the conversion factors as well as some indication of the magnitude of the different energy units.

Energy Units

The energy content of fuels and electricity is expressed in different units around the world. The units that are used in this book, and their equivalence to other energy units, are as follows:

quad (quadrillion Btu) = 1.055 EJ (exajoules)
ton of oil equivalent (toe) = 41.9 GJ (gigajoules) = 39.7 million Btu
barrel of oil (bbl) = 6.1 GJ (gigajoules) = 5.8 million Btu
ton of coal equivalent (tce) = 29.3 GJ (gigajoules) = 27.8 million Btu
kilowatt-hour (kWh) = 3.6 MJ (megajoules) = 3,412 Btu
watt = 1 joule per second = 3.412 Btu per hour

To put these energy units in perspective, here are some typical energy consumption levels:

- A typical U.S. household consumes about 180 million Btu (190 GJ) of energy per year (including energy losses in electricity production and delivery).
- A typical U.S. car or light truck consumes about 600 gallons of gasoline per year, equivalent to 75 million Btu (79 GJ) of energy.
- A typical rural household in a developing country consumes about 38 million Btu (40 GJ) of traditional fuels (wood, crop residues, and animal wastes) per year.
- A 60-watt light bulb used 4 hours per day consumes 88 kWh per year.
- A typical U.S. refrigerator now in use consumes about 900 kWh per year.

The term *primary energy* is used at times in this book. Primary energy includes losses in the production and delivery of fuels and electricity.

In addition, the following metric prefixes are used in this book:

kilo (k) – 10^3 (thousand)
mega (M) – 10^6 (million)
giga (G) – 10^9 (billion)
tera (T) – 10^{12} (trillion)
peta (P) – 10^{15} (quadrillion)
exa (E) – 10^{18}

Acknowledgments

I am indebted to many individuals and organizations who assisted me during the journey that culminated in this book. First, the book builds on a number of studies I conducted or participated in while working at the American Council for an Energy-Efficient Economy (ACEEE) in Washington, D.C. Chapter 5 in particular draws from previous collaborations and the critical input provided especially by Steven Nadel and also John DeCicco, Neal Elliott, Toru Kubo, and Jennifer Thorne of ACEEE. The integrated analysis in Chapter 5 was provided by Steve Bernow, Bill Dougherty, and Allison Bailie of the Tellus Institute in Boston, Massachusetts. I greatly appreciate their excellent work on this project as well as other joint studies over the years. In addition, Eric Kemp-Benedict of the Tellus Institute provided valuable assistance as I was completing the energy scenario in Chapter 8.

Chapter 6 was a collaborative effort with long-time colleagues in Brazil, namely Roberto Schaeffer, Alexandre Szklo, and Mauricio Tolmasquim from the Federal University of Rio de Janeiro. Their help was invaluable. I also thank other Brazilian colleagues who helped (knowingly or unknowingly) with portions of this book including José Alencar, José Goldemberg, Paulo Leonelli, Marcos Lima, Regiane Monteiro de Abreu, Geraldo Pimentel, and Paulo Cezar Tavares.

This book would not have been possible without the support and insight from Todd Baldwin, my editor at Island Press. I believe the book greatly benefited from his suggestions on both missing links and how best to fit the links together. Todd also provided valuable guidance on communicating to a broad audience. I also want to thank Becky Campbell-Howe for her very capable editorial assistance, Renee Nida for her help in editing early drafts of some of the chapters, and James Nuzum for his assistance at Island Press.

Many individuals provided useful information and helpful comments on portions of the manuscript including Nils Borg, Shivan Kartha, Roberto Lamberts, Benoit Lebot, Eric Martinot, Alan Miller, Isac Roizenblatt, Art Rosenfeld, Roberto Schaeffer, and Michael Shepard. I especially thank José Roberto Moreira for carefully reviewing nearly all of the chapters in draft form.

I thank the Energy Foundation of San Francisco, in particular Hal Harvey (now with the Hewlett Foundation), Eric Heitz, and Polly Shaw, for funding this book directly as well as indirectly through various grants over the years. This book would not have been possible without their generous support. I also appreciate and thank the United Nations Development Programme for requesting and supporting a study that contained early versions of Chapters 3 and 4, and the U.S. Environmental Protection Agency for funding a study that led to Chapter 6. In addition, I want to thank the Land and Water Fund of the Rockies of Boulder, Colorado, for their support as I started the Southwest Energy Efficiency Project and completed this book in late 2001 and 2002. Boulder turned out to be an ideal place to hunker down and get it done.

Finally, I would like to thank my family—my wife Luci, daughter Sara, and son Will—for their encouragement and support. I spent many evenings and weekends working on this book, and I greatly appreciated their patience and understanding.

While I received valuable assistance from all these individuals and organizations, I am responsible for any errors or omissions remaining in the book. I would appreciate hearing about these, or receiving other feedback on the book, via e-mail to hgeller@swenergy.org.

Howard Geller
Boulder, Colorado
June 2002

Introduction

Energy affects nearly every aspect of our lives. We need energy to heat, cool, and light our homes as well as to cook and refrigerate our food. Energy fuels our cars, trucks, and other means of transport. Energy powers our industries, farms, offices, and other workplaces. In the United States and other industrialized nations, nearly all of this energy is derived from fossil fuels (oil, coal, and natural gas) and electricity.

Fossil fuels and electricity appear to be plentiful, inexpensive, and readily available. Ignoring taxes, a gallon of gasoline costs about as much as a gallon of bottled water. We turn on our appliances and lights with the flip of a switch, giving little thought to how the electricity is generated, or the consequences of its generation. We fill our gasoline tanks with little thought about where the fuel comes from, or the consequences of our fossil fuel–intensive culture.

Energy reaches into our lives in other less direct ways as well. Energy producers such as the major oil companies are among the largest and most profitable corporations in the world. The actions of these companies affect governments and the world economy, as witnessed by the spectacular collapse of Enron. The distribution and pursuit of energy resources worldwide affect relations among nations, as witnessed by the periodic conflicts over oil in the Persian Gulf region or the struggles between the Organization of Petroleum Exporting Countries (OPEC) and oil-importing nations (Yergin 1991).

The first theme of this book is that current energy sources and patterns of energy use are unsustainable. Continuing to consume ever greater amounts of fossil fuels will cause too much damage to the environment, risk unprecedented climate change, and rapidly deplete petroleum resources. Current trends in energy supply and demand will also exacerbate inequity and tensions among nations, tensions that fuel regional conflict and outbursts reminiscent of the attacks on the World Trade Center and Pentagon. In short, continuing on a business-as-usual energy path will put the well-being of future generations at risk.

The second theme is that an "energy revolution" is possible and desirable. By emphasizing much greater energy efficiency and growing reliance on renewable energy sources such as solar energy, wind power, and bioenergy, all of the problems associated with current energy patterns and trends can be mitigated. However, a formidable set of barriers is limiting the rate of energy efficiency improvement and the transition to renewable energy sources in most parts of the world.

The third and primary theme of this book is that it is possible to overcome these barriers through enlightened public policies. Experience with policies for increasing energy efficiency and renewable energy use is growing, providing many successful models and lessons that can guide further action. Expanding the adoption of successful policies as well as increasing and focusing international efforts could accelerate the energy revolution and result in a more sustainable energy future.

Before considering future energy policies and scenarios, it is useful to review global energy use over the past century or two. Worldwide energy use has increased 20-fold since 1850, 10-fold since 1900, and more than 4-fold since 1950 (Fig. 1-1). This dramatic increase in energy use enabled economic growth and higher standards of living for a sizable fraction, but not all, of the world's growing population. Most of the growth in energy use during the past 100 years occurred in the industrialized world, home to about 20 percent of the world's population.

Our sources and uses of energy have changed a great deal over the past 150 years. Most energy consumed in the nineteenth century was in the form of biomass—fuelwood, charcoal, and agricultural residues—also known as traditional energy sources. Coal production expanded rapidly in the latter half of the nineteenth century, making coal the dominant energy source worldwide for about 75 years starting around 1890. Use of coal in steam engines and as a fuel for electricity generation transformed industries and life, at least in more developed countries. Petroleum production rapidly accelerated following World War

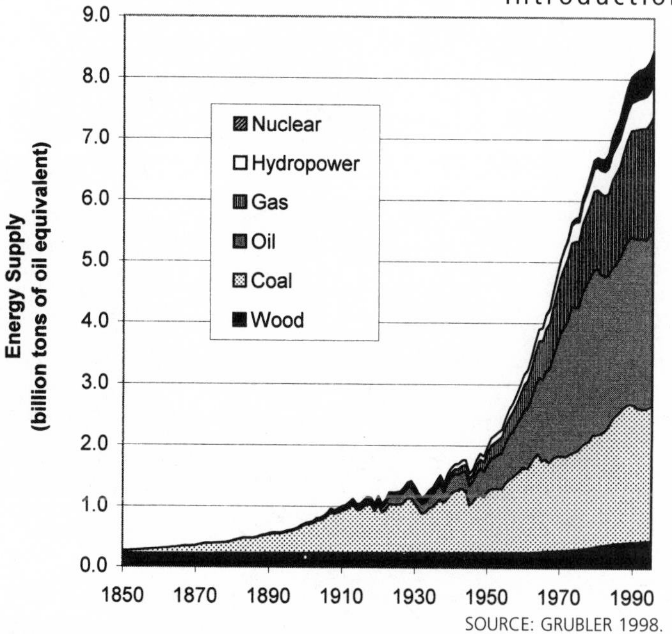

FIGURE 1-1: World energy supply since 1850.

II, making petroleum the dominant energy source over the past 35 to 40 years. Use of petroleum products in cars, buses, trucks, aircraft, and other types of vehicles transformed mobility. In addition, use of natural gas and nuclear power both grew rapidly during the past 30 years. Thus, the world has experienced energy transitions before, and these transitions coincided with and literally fueled economic and social transitions (Grubler 1998).

Fossil fuels provide about 80 percent of the global energy supply today. Among fossil fuels, petroleum provides the largest share and accounts for about 35 percent of the total global energy supply. Coal provides about 23 percent and natural gas about 21 percent of the total. Renewable energy sources account for about 14 percent of the global energy supply, but most of this is in the form of traditional energy sources.[1] Modern renewable energy sources, including hydropower, wind power, and modern forms of bioenergy, account for only about one-third of the renewable energy total. Nuclear energy provides the remaining 6 percent of the global energy supply (UNDP 2000).

About one-third of the world's population—2 billion people—still rely almost entirely on firewood and other traditional energy sources for their energy needs. These households do not consume electricity, petroleum products, or natural gas—a major factor contributing to their impoverishment. Meanwhile, wealthier citizens of the world consume increasing amounts of fossil fuels, hydropower, and nuclear energy to power ever-larger vehicles, buildings, and appliances.

Current Energy Trends and Their Implications

Business-as-usual forecasts project that global energy use will increase around 2 percent per year in the coming decades. For example, the 2000 World Energy Outlook produced by the International Energy Agency projects in its *Reference Scenario* that world energy demand will increase by 54 percent between 1997 and 2020 (Fig. 1-2) (IEA 2000a). Oil use would increase by 56 percent, natural gas use by 86 percent, and coal use by 49 percent in this forecast. Fossil fuels would account for nearly 84 percent of total primary energy supply in 2020, up from their 80 percent share in 1997. Use of traditional fuels in developing nations would continue to increase, but more slowly than the projected growth in fossil fuel use.

Other forecasts indicate that if current energy policies and trends continue, global energy use could double from the level in 1990 by about 2025, triple by 2050, and further rise in the latter half of the twenty-first century (Nakicenovic, Grubler, and McDonald 1998). The majority of this growth is expected to take place in developing countries given their high population growth and low levels of energy consumption at the present time. Developing countries could pass industrialized countries in terms of total energy use by around 2025. But per

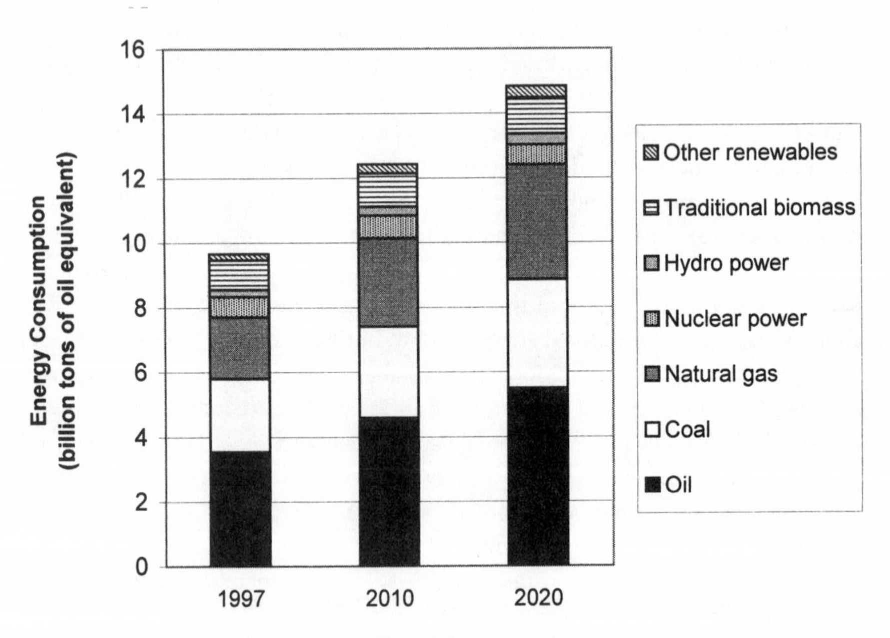

SOURCE: IEA 2000a.

FIGURE 1-2: Reference scenario of world primary energy consumption.

capita energy use in industrialized nations still would increase and would remain far higher than in developing countries in these business-as-usual forecasts.

A high-growth, fossil fuel–intensive energy future presents a variety of problems and challenges for humanity. These problems and challenges include high costs, air pollution, global warming, security risks, resource depletion, and inequity.

High Costs

Building power plants, oil and gas pipelines, and other energy supply facilities is very capital-intensive. Analyses show that if worldwide energy use continues to rise on the order of 2 percent per year, energy supply investments of $11 to $13 trillion will be needed during the period from 2000 to 2020 and an additional $26 to $35 trillion will be needed from 2020 to 2050 (in 1998 dollars).[2] This level of investment—$500 billion to $1 trillion per year—is two to four times the level of investment in energy production and conversion worldwide during the 1990s. Investment in energy supply could grow to 7 to 9 percent of the gross domestic product (GDP) in the transition economies over the next 20 years, for example (Nakicenovic 2000).

Expanding investment in energy supply and conversion is feasible in some countries, but will be difficult in transition and developing nations. These countries need to invest in a broad range of priorities including education, sanitation, health care, and rural development. Many developing and transition countries are strapped with high debts and have a difficult time attracting investment from the private sector. These factors limit investment in energy supply in Asia and Africa, for example, which in turn inhibits social and economic development there (Rogner and Popescu 2000).

Energy costs are also felt on the individual level in developing and transition countries. Households in developing countries often pay a sizable portion of their income for energy, including kerosene, batteries, and other fuels, and often use these energy sources very inefficiently. In former communist nations, households face very high energy bills relative to their diminished income because of inefficiency and energy waste, along with the phasing out of energy price subsidies. In Ukraine, for example, energy costs account for as much as 40 percent of household expenditures (IEA 2001g).

Likewise, energy expenditures can "eat up" a sizable fraction of the income of poorer families in industrialized countries, due in large part to poorer families living in inefficient homes. In the United States, for example, energy costs equal 12 to 26 percent of the income of poorer households compared to just a few percent of income for middle-income and wealthier households (NCLC 1995). A significant fraction of the world's population will continue to waste energy and face high energy costs under a business-as-usual energy future.

Local and Regional Air Pollution

Burning fossil fuels causes air pollution that is harming public health and disrupting ecosystems. Energy activities account for 85 percent of humanmade emissions of sulfur dioxide (SO_2), 45 percent of particulate emissions, 41 percent of lead emissions, 40 percent of hydrocarbon emissions, and 20 percent of nitrous oxide emissions to the atmosphere (Holdren and Smith 2000). These air pollutants in turn result in acid rain, urban smog, and hazardous soot. Burning fossil fuels is also a major source of toxic chemicals that are known to cause cancer (EPA 2002a).

It is estimated that 1.4 billion people are exposed to dangerous levels of outdoor pollution (Watson et al. 1998). Due to poor combustion efficiencies and lack of pollution controls, levels of particulate matter are two to five times higher than World Health Organization (WHO) limits in Southeast Asian cities and even higher in some cities in China and India (Li 1999). Over 80 percent of major cities in China exceed WHO limits for SO_2, some by up to a factor of three (Li 1999). Levels of lead, carbon monoxide, nitrogen oxides, and volatile organic compounds also often exceed safe levels.

The impacts of this air pollution are stark. Outdoor air pollution is especially high in urban areas, causing on the order of 500,000 deaths globally and up to 5 percent of deaths in urban areas in some developing countries (WHO 1997). It is estimated that urban air pollution causes 170,000 to 290,000 premature deaths annually in China and 90,000 to 200,000 in India (Holdren and Smith 2000). The human health effects of air pollution in Chinese cities, when translated into economic terms, exceed 20 percent of the average income of Chinese workers and are approaching $50 billion (7 percent of GDP) for the nation as a whole (World Bank 1997).

Air pollution from burning fossil fuels is not just a problem in developing countries. It is estimated that power plant emissions cause about $70 billion of harm to human health, buildings, and crops in the European Union annually (Krewitt et al. 1999). This is equivalent to $0.045 per kilowatt-hour (kWh), about half the average retail electricity price. It is also equivalent to about 1 percent of the European Union's GDP. Most of these costs are due to adverse impacts on public health. For example, it is estimated that air pollution causes about 800,000 episodes of asthma and bronchitis and 40,000 deaths annually in Austria, France, and Switzerland (London and Romieu 2000).

The emissions of most air pollutants declined in the United States over the past 20 years. But air pollution, especially elevated ozone and particulates levels, is still a problem in many metropolitan areas. Approximately 125 million Americans (46% of the total U.S. population) live in counties that did not meet the air quality standards for at least one pollutant in 1999 (ALA 2001). Hundreds of thousands of Americans suffer from asthma attacks and other respiratory problems due to fine particulate emissions from power plants and other sources. Long-term exposure to these tiny particles results in increased risk of lung cancer and heart disease, and cuts short the lives of over 30,000 people in the United States each year (Clean Air Task Force 2000).

Environmental contamination can be especially bad in regions with high levels of energy production. Kazakhstan, for example, is a major producer of oil, natural gas, coal, and uranium. But it also has severe air pollution, soil contamination, and pollution of both surface and ground water (Dahl and Kuralbayeva 2001). Pollution has severely affected the Caspian Sea and its ecosystems. In addition, the uranium and fossil fuel industries have contributed to radioactive contamination on a wide scale. In short, Kazakhstan faces a public health and ecological crisis due to energy-related pollution.

As bad as outdoor air pollution is in many developing countries, indoor air pollution from burning fuelwood and agricultural residues for cooking and heating is an even greater health hazard. In South Africa, for example, rural households that burn wood for cooking and heating exhibit indoor particulate levels 13 times the maximum level recommended by the WHO. Epidemiological research shows that individuals exposed to this level of particulate matter have five times the risk of contracting respiratory illness compared to those living under normal conditions (Spalding-Fecher, Williams, and van Horen 2000). Households in South Africa that use coal for heating and cooking are also exposed to dangerous levels of particulate matter.

According to the WHO and other experts, indoor air pollution is causing about 1.8 million premature deaths annually worldwide, mainly in women and children (WHO 1997). This is three to four times greater than the deaths caused by outdoor air pollution worldwide. In India alone, indoor air pollution from burning solid fuels causes about 500,000 premature deaths per year in women and children (Holdren and Smith 2000). This is greater than the mortality caused by other major health hazards in India such as malaria, acquired immune deficiency syndrome (AIDS), heart disease, and cancer.

Fossil fuel–intensive energy development over the next century could exacerbate these air quality problems, adversely affecting economic output as well as public health. With growing use of fossil fuels and limited pollution abatement, ambient air quality could further deteriorate and severely affect public health, food production, and ecosystems in Asia within 20 years (Nakicenovic, Grubler, and McDonald 1998). Also, business-as-usual energy development envisions that billions of people will continue to burn fuelwood and other traditional fuels for cooking and heating, with continued high incidences of respiratory disease and premature death as a result.

Global Warming

Carbon dioxide and other "greenhouse gases" are building up rapidly in the atmosphere and causing global warming. There has been a 31 percent rise in atmospheric carbon dioxide levels and a 151 percent rise in methane levels since preindustrial times. The carbon dioxide concentration is higher today than at any point during the past 420,000 years, and the rate of increase is unprecedented during at least the past 20,000 years (IPCC 2001a).

With this buildup of carbon dioxide and other greenhouse gases, the average temperature of the earth's surface increased about 1.1°F (0.6°C) over the past century (Fig. 1-3) (IPCC 2001a). Furthermore, the 1990s were the warmest decade on record; 1998 was the single warmest year of the past 1,000 years, and 2001 was the second warmest year (ENS 2001c).

Energy-related activities, mainly burning of fossil fuels, produce about 78 percent of humanmade carbon dioxide emissions and about 23 percent of humanmade methane emissions (Holdren and Smith 2000). Carbon dioxide and methane are responsible for about 80 percent of the warming that has occurred since preindustrial times due to emissions of long-lived gases. In the United States, carbon dioxide accounts for nearly 85 percent of the greenhouse gas emissions addressed by the United Nations Climate Change Convention (see Table 1-1). Energy-related activities also result in emissions of sulphates and particulates as well as ozone formation in the troposphere, all of which have an effect on global warming. Given the increase in temperature over the past century and decade in particular, scientists now conclude that human activities have caused most of this warming (IPCC 2001a).

Global warming is starting to have a variety of adverse impacts. These include more frequent and extreme weather events such as droughts, floods, and heat waves, which in turn cause death, property damage, and crop loss. Worldwide, economic losses due to extreme weather events increased 10-fold from about $4 billion per year during the 1950s to about $40 billion per year during the 1990s (IPCC 2001b). Global warming also raises the sea level, which adversely affects

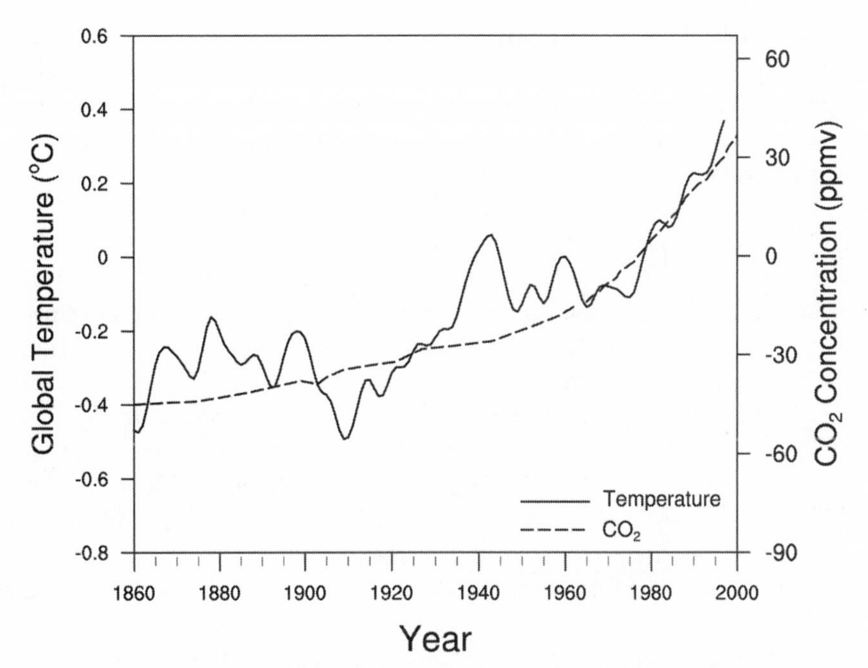

NOTE: ppmv = parts per million by volume.
SOURCE: HURRELL 2002.

FIGURE 1-3: Trends in the global mean surface temperature and atmospheric carbon dioxide concentration during 1860–2000 (changes from the average during 1961–1990).

TABLE 1-1

Recent Trends in U.S. Greenhouse Gas Emissions

	Annual Emissions Million metric tons of carbon equivalent		
Gas	1990	1995	2000
Carbon dioxide	1363	1447	1593
Methane	178	179	168
Nitrous oxide	106	114	116
HFCs, PFCs, and SF_6	26	27	33
TOTAL	1672	1768	1909

NOTE: HFCs = hydrofluorocarbons, PFCs = perfluorocarbons, SF_6 = sulfur hexafluoride.
SOURCE: EPA 2002b.

low-lying island nations and coastal areas. The average sea level rose 0.1 to 0.2 meters during the twentieth century. In addition, global warming is causing glaciers to retreat and the polar ice caps to shrink, and is thought to be harming other vulnerable natural systems such as coral reefs, atolls, mangroves, and boreal and tropical forests (IPCC 2001b, DOS 2002).

If current energy supply and demand trends continue, dramatic global warming will occur during the twenty-first century. Worldwide carbon dioxide emissions would increase by a factor of 2 to 2.5 by 2050 and 2.5 to 3.5 by 2100 assuming a high-growth, fossil fuel–intensive future. As carbon dioxide emissions increase, oceans and land will absorb a smaller fraction of emissions. Thus the buildup of carbon dioxide emissions in the atmosphere will accelerate, leading to concentrations of 700 to 970 parts per million by volume (ppmv) by 2100 (IPCC 2001a, Nakicenovic, Grubler, and McDonald 1998). This carbon dioxide concentration is 2.5 to 3.5 times the preindustrial level. Along with the buildup of other greenhouse gases, it would lead to a 2.5 to 10°F (1.4–5.8°C) average surface temperature rise by 2100 according to the latest estimates (IPCC 2001a).

A temperature rise even at the lower end of this range could have devastating effects. It would greatly increase the frequency and magnitude of severe weather events such as droughts, hurricanes, and tornados; spread infectious diseases such as malaria and dengue fever; increase mortality from heat waves; reduce crop yields in most inhabited regions; and damage ecosystems worldwide (IPCC 2001b). It would affect rain and snowfall, for example, diminishing snowfall in the western United States, which in turn would reduce water supply in this semiarid region (DOS 2002). It could cause a sea level rise of up to 90 centimeters this century, greatly increasing flooding of low-lying areas and inundating island nations. This in turn would displace tens of millions of people. The economic cost of these various adverse impacts could reach $300 billion annually by 2050, equivalent to 1.5 percent of global GDP, according to a major insurance firm (ENS 2001a). Poor countries would be most severely affected since they are highly vulnerable to these impacts (e.g., crop yields are expected to decline in most tropical and subtropical regions). Furthermore, developing countries have limited resources to adapt to climate change.

Projected global warming during the twenty-first century based on business-as-usual energy trends also could cause catastrophic and irreversible ecological changes. These potential changes include a significant slowing or even shutdown of ocean circulation that transports warm water to the North Atlantic, collapse of the West Antarctic ice sheet leading to a sea level rise of many meters, and possible "runaway global warming" due to carbon cycle feedbacks, release of carbon from permafrost regions, and/or release of methane from hydrates in coastal sediments (Holdren and Smith 2000). The likelihood of these changes is still not well-known and is probably very low; but the likelihood will increase if carbon dioxide rapidly builds up in the atmosphere and temperatures significantly rise during and beyond the twenty-first century.

Even if these catastrophic events do not occur, increasing carbon dioxide emissions and sustained global warming inevitably would result in both melting of the polar ice sheets and ocean thermal expansion over the course of many centuries. This in turn would result in sea level rise of at least a number of meters during the next thousand years or so. The impact is delayed because of the time lags between global warming and sea level rise, but is "wired in" by elevated temperature levels in the nearer term (Mahlman 2001). Sea level rise of this magnitude would inundate vast areas and present enormous challenges to humankind.

Security Risks

The United States and other industrialized nations are highly dependent on imported oil, and this dependence is increasing. This leaves Western economies vulnerable to price fixing by the OPEC cartel and potential oil price shocks. It also poses national security risks due to potential oil supply disruptions, military intervention to maintain vital oil supplies, and the side effects of a strong military presence in the Persian Gulf region. In fact there have been 14 significant oil supply disruptions during the past 50 years, mostly related to political or military conflict in the Middle East (IEA 2001h).

Previous oil price shocks cost the United States trillions of dollars due to subsequent inflation, recession, and transfer of wealth to countries that maintained oligopolistic control over energy prices (Greene and Leiby 1993). Western nations spend tens of billions of dollars each year protecting oil supplies in the Middle East (UCS 2002). And several hundred billion dollars were spent on the 1990–91 war in Iraq, a war fought in large part to protect oil supplies (Khatib 2000).

Organization for Economic Cooperation and Development (OECD) nations are expected to increase their oil import dependence from 54 percent in 1997 to 70 percent in 2020 if current trends are maintained (IEA 2000a).[3] Also, Asian nations are expected to greatly increase their dependence on imported oil over the next 20 years. The share of world oil supply coming from OPEC nations in the Persian Gulf region increases from 26 percent in 1997 to 41 percent in 2020 in the 2000 World Energy Outlook *Reference Scenario* forecast. Given the existence of the OPEC cartel and the political instability in the Persian Gulf region, oil-importing nations will face even greater economic and security risks in the future if this scenario plays out.

The terrorist attacks on the World Trade Center and Pentagon were not unconnected to our high oil import dependence. Revenues from oil sales finance terrorist groups such as the al-Qaeda network and the regime of Saddam Hussein in Iraq. Osama bin Laden was enraged by the presence of American soldiers in Saudi Arabia, a presence driven by our thirst for oil and concern for stability in the region (Krugman 2001). The fact that Western nations support undemocratic, repressive governments in the Middle East as long as they keep the oil flowing also contributes to poverty, frustration, and terrorism.

The world oil supply infrastructure itself is vulnerable to terrorist attack and disruption. Particularly vulnerable elements include major oil pipelines such as the Trans-Alaska pipeline, tanker terminals, and the large amount of oil flowing through the narrow Persian Gulf (Banerjee 2001). In fact, oil supply through the Trans-Alaska pipeline has been disrupted many times. For example, the pipeline was shut down for nearly three days shortly after the September 11, 2001, terrorist attacks when an intoxicated man shot holes in it, causing nearly 7,000 barrels of oil to leak out before the spill was detected (ENS 2001b).

Key components of our electricity supply network, including nuclear power plants and major transmission lines, are also vulnerable to attack. A bomb exploding at or airliner crashing into a nuclear power plant or spent nuclear fuel storage site, for example, would release large quantities of radioactive isotopes. This release could kill or injure tens of thousands of people living downwind of the plant (ENS 2001d, UCS 2002). In addition, nuclear energy technologies, such as uranium enrichment and plutonium reprocessing activities, could contribute to nuclear weapons proliferation through either state-sponsored efforts or diversion of nuclear weapons materials by individuals (Williams 2000).

Resource Depletion

U.S. oil production peaked in 1970 and declined nearly 30 percent by 2000. Put simply, the United States is running out of economically recoverable oil. This is true whether or not oil drilling is allowed in the Arctic National Wildlife Refuge (ANWR) and other environmentally sensitive areas (Geller 2001).

World oil production will follow a similar path, peaking at some point in the next few decades and then declining due to resource depletion. There is controversy, however, regarding whether production will peak sooner (i.e., within 10 years) or later (i.e, in a few decades). The U.S. Geological Survey has produced optimistic estimates of 3 trillion barrels of ultimate recoverable oil reserves in the world. With this total amount and the assumption that world oil demand

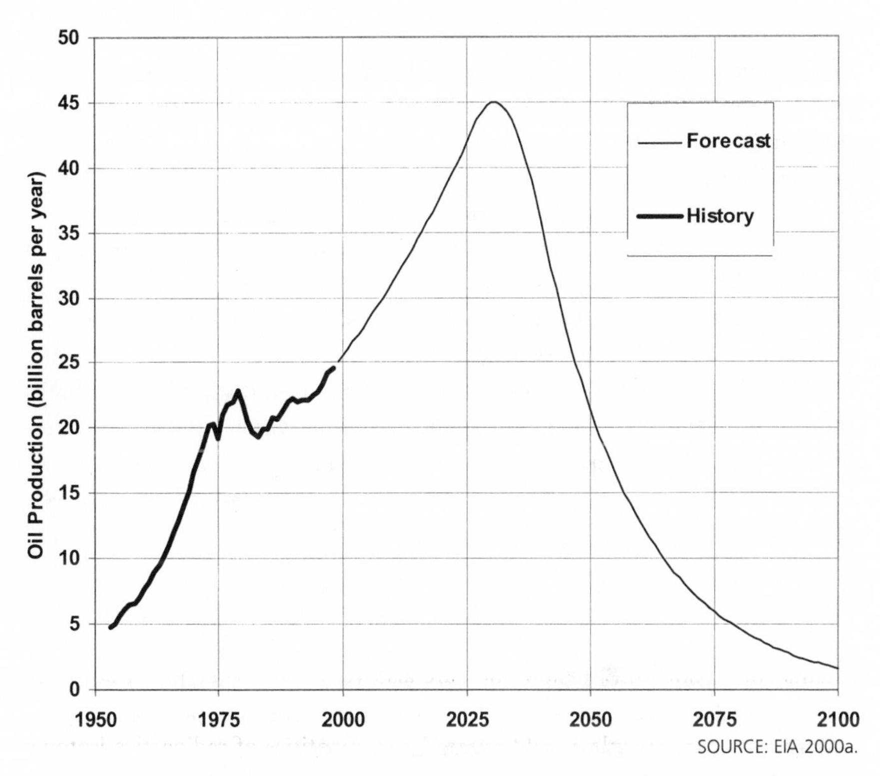

SOURCE: EIA 2000a.

FIGURE 1-4: World oil production forecast based on USGS estimate of 3.0 trillion barrels of ultimate oil recovery.

continues to rise 2 percent per year, oil production peaks around 2030 (Fig. 1-4) (EIA 2000a). Other experts believe that much less oil will ultimately be recovered, perhaps only around 2 trillion barrels (Campbell and Laherrere 1998, Deffeyes 2001). This value leads to world oil production peaking by 2010. The steady decline in oil discoveries throughout the world over the past 30 years suggests that the latter view is more plausible (Bentley 2002).

Once conventional oil production peaks, it is expected to decline around 3 percent per year as remaining resources become increasingly difficult to reach (Bentley 2002). Oil prices could climb steeply as production declines, unless alternative fuels become available in large quantity.

There are vast reserves of unconventional oil resources such as oil shale and tar sands in the world (Rogner 2000). Many oil companies are pursuing these resources since they are natural extensions of conventional oil production. But developing these resources on a large scale would be costly, result in massive environmental damage, and contribute heavily to global warming. The contribution to global warming would be high because obtaining, processing, and burning these resources would result in more greenhouse gas emissions per unit of energy than producing and burning conventional oil (Campbell and Laherrere 1998, Williams 2000). Thus, unconventional oil resources are not an attractive alternative fuel option.

Resource depletion is much less of an issue for coal and natural gas than for oil. Coal reserves and likely future additions to reserves would last hundreds of years at current production rates (Rogner 2000). The nearer-term constraint on coal production and use is environmental, primarily global warming. Likewise, conventional natural gas reserves, along with enhanced gas recovery techniques, would last hundreds of years at current production rates (Rogner 2000). In addition, there are enormous quantities of unconventional gas in aquifers, polar permafrost, and below the ocean floor. Although there are no economically viable techniques for recovering these resources today, such techniques may be developed in the future.

Inequity

Energy consumption, like income, is distributed very inequitably around the world. Table 1-2 shows per capita commercial energy consumption trends by region of the world. Per capita energy consumption of commercial fuels and electricity is growing most rapidly in developing countries, but the OECD nations still consumed about six times more commercial energy per capita than developing nations as of 1997. Per capita consumption in North America was 10 times that in Asia and Africa (UNDP 2000).

There are also substantial disparities in energy use within countries, both in industrialized and developing countries. In the United States, for example, wealthier households consume about 75 percent more energy than poorer households. Likewise, urban households in China's four richest coastal provinces consume about two and a half times as much energy as households in poorer interior provinces. Similar differences are found in other countries, including Brazil, India, and Mexico (Smil 2000).

Electricity consumption is even less equitably distributed than commercial energy as a whole. Per capita electricity use in the OECD nations was 13 times that in East Asia, nearly 65 times that in sub-Saharan Africa (excluding South Africa), and 26 times that in South Asia as of 1996 (UNDP 2000). Per capita electricity use is rising in Asia, but not in sub-Saharan Africa.

Given this inequitable distribution of commercial energy consumption, it is not surprising that greenhouse gas emissions and contributions to global warm-

TABLE 1-2

Distribution of Commercial Energy Use (gigajoules per capita)

Region	1971	1980	1990	1997	Avg. growth 1971–97 (%/yr)
North America	266	276	263	272	0.1
Latin America	36	42	40	47	1.0
OECD Europe	118	134	137	141	0.7
Non-OECD Europe	76	108	108	84	0.4
Former Soviet Union	135	178	195	129	0.2
Middle East	35	61	77	95	3.9
Africa	23	26	27	27	0.6
China	20	25	32	38	2.5
Asia other than China	15	17	21	26	2.1
Pacific OECD[a]	94	113	142	174	2.4
TOTAL	62	69	70	70	0.5

NOTES: [a]Includes Republic of Korea. OECD = Organization for Economic Cooperation and Development.
SOURCE: UNDP 2000.

ing are highly skewed as well. In 1995, the 20 percent of the world's population in countries with the highest per capita energy use contributed 63 percent of total carbon dioxide emissions from burning fossil fuels. In contrast, the 20 percent of the world's population in countries with the lowest per capita energy use contributed just 2 percent of emissions (UNFPA 2001).

Conventional energy policies and trends emphasize increasing energy use among the wealthier citizens of the world (in both industrialized and developing nations), rather than providing modern energy sources and improved energy services to poorer citizens. Nearly half the world's population lives in rural areas of developing countries, and most of these people do not use electricity or modern cooking fuels (Goldemberg 2000). In India, for example, less than 30 percent of rural households use electricity, over 90 percent use traditional biomass cooking fuels, and more than 55 percent of farmland is cultivated by animal power (Pachauri and Sharma 1999). In many African nations, less than 20 percent of the population has access to electricity (Balu 1997). Per capita electricity use in sub-Saharan Africa (excluding South Africa) is only about 126 kWh, only 1 to 2 percent that in North America or Western Europe (Karekezi and Kimani 2002).

While hundreds of millions of rural households in developing countries began using electricity and improved their cooking methods between 1970 and 1990, about 2 billion people were without access to electricity or improved cooking methods in 1990—about the same number as in 1970 (Goldemberg 2000). Women are especially impacted by the lack of modern energy sources in much of the developing world since they are the main gatherers and users of fuelwood and other traditional fuels (Cecelski 1995). Business-as-usual scenarios do not envision this pattern changing significantly in the next few decades (IEA 2000a).

Conventional energy policies emphasize a centralized approach to rural electrification and fuels distribution. This approach is high cost and often fails to serve the needs of poorer households who cannot afford to hook up to the electric grid or to purchase conventional cooking fuels such as bottled gas (Goldemberg 2000).

The Energy Revolution—Toward a Sustainable Future

Sustainable energy development should provide adequate energy services for satisfying basic human needs, improving social welfare, and achieving economic development throughout the world. Sustainable energy development should not endanger the quality of life of both current and future generations, and it should not threaten critical ecosystems (Rogner and Popescu 2000). For the reasons described above, current energy patterns and trends will not result in a sustainable energy future.

A sustainable energy future is possible through much greater energy efficiency and much greater reliance on renewable energy sources compared to current energy patterns and trends. Greater energy efficiency would reduce growth in energy consumption, decrease investment requirements, and improve energy services in poorer households and nations. Shifting from fossil fuels to renewable energy sources in the coming decades would address all the problems associated with a business-as-usual energy future.

This type of energy future is illustrated by the ecologically driven Low Growth, Low Carbon Scenario developed in a study by the International Institute for Applied Systems Analysis (IIASA) and the World Energy Council (WEC) (Fig. 1-5). In this scenario, accelerated energy efficiency improvements limit the growth in global energy use to 0.8 percent per year during the next century. Likewise, modern biomass, solar, and other renewable sources start to contribute a significant fraction of global energy supply within two decades. Renewable energy sources provide 40 percent of total energy supply by 2050 and 80 percent by the end of the twenty-first century (Nakicenovic, Grubler, and McDonald 1998).

The Low Growth, Low Carbon Scenario leads to nearly 65 percent less carbon emissions by 2050 and over 90 percent less carbon by 2100 compared to the High Growth, High Coal Use Scenario developed by IIASA-WEC (Nakicenovic, Grubler, and McDonald 1998). The Low Growth, Low Carbon Scenario would

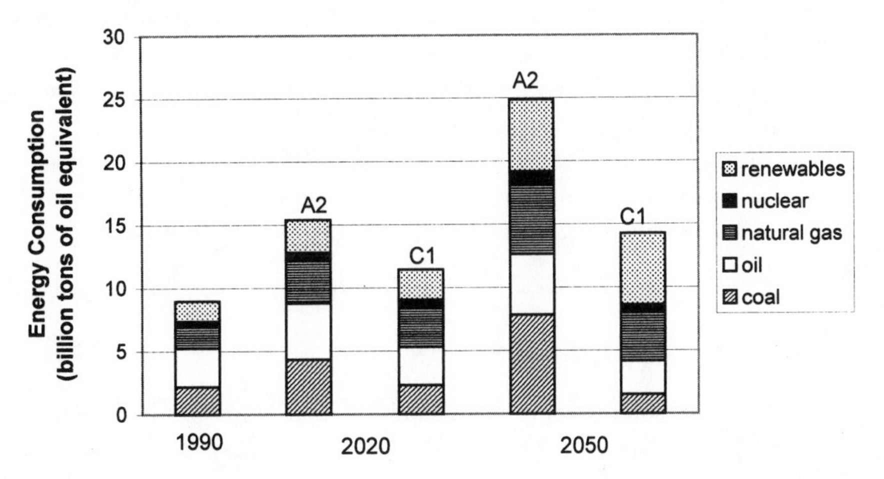

SOURCE: Nakicenovic, Grubler, and McDonald 1998.

FIGURE 1-5: Comparison of global energy consumption in the IIASA–WEC High Growth, High Coal Scenario (A2) and the Low Growth, Low Carbon Scenario (C1).

meet the greenhouse gas emissions targets set in the Kyoto Protocol and then stabilize atmospheric carbon dioxide at about 450 ppmv in the latter part of the twenty-first century. This is sufficient to greatly reduce the risks associated with global warming.

Reducing total energy use and increasing the share provided by renewable sources would reduce urban air pollution, acid rain, and other regional and local environmental problems. SO_2 emissions in 2050 are about two-thirds less in the Low Growth, Low Carbon Scenario compared to the High Growth, High Coal Use Scenario, for example. An emphasis on energy efficiency and renewable energy sources also reduces total investment requirements by 33 to 50 percent (Nakicenovic, Grubler, and McDonald 1998), and it would lead to much improved energy services to poorer citizens and rural areas of developing nations.

Similarly, the IPCC Third Assessment Report includes a set of scenarios that assume vigorous improvements in resource efficiency, reductions in materials intensity, and emphasis on sustainability worldwide. These assumptions lead to carbon dioxide concentrations of 450 to 550 ppmv by 2100 and an estimated temperature rise around 2.5 to 3.6°F (1.4–2.0°C) over the time period from 1990 to 2100 (IPCC 2001a, 2001c). In contrast, the high demand growth, fossil fuel–intensive scenarios developed by the IPCC lead to carbon dioxide concentrations of 700 to 970 ppmv by 2100, an estimated temperature rise of up to

10.4°F (5.8°C) from 1990 to 2010, and much greater local and regional air pollution (IPCC 2001a, 2001c).

This book's unique focus is on the policies for achieving an energy revolution. But in considering these policies, it is helpful to have an understanding of the key technologies that would be employed in the revolution as well as the context in which these technologies and policies will operate during the twenty-first century. The following brief reviews are intended to provide this background and context.

Renewable Energy Sources

Renewable energy sources could provide all of the energy consumed in the world. There is more than enough solar energy potential alone to meet projected global energy needs well beyond 2100 (Rogner 2000). In addition, the feasible wind power potential onshore is estimated to be 20 to 50 trillion kWhs per year—1.5 to 4 times current worldwide electricity production (Turkenburg 2000). Adding offshore wind resources would increase this potential. And biomass energy plantations, considering expected land availability in the future, could provide the equivalent of one-quarter to three-quarters of current worldwide energy consumption (Turkenburg 2000).

Table 1-3 shows the status of the major renewable energy technologies. The performance and cost of these technologies has dramatically improved over the past two decades (Interlaboratory Working Group 1997, Turkenburg 2000). Some newer renewable energy technologies are now competitive with conventional energy sources in specific applications. Wind power, for example, is very close to being competitive with electricity from new fossil fuel–based power plants. Solar photovoltaic (PV) power and modern forms of biomass can be economical in rural areas of developing countries where grid-based electricity and fossil fuels are either costly or not available. Taking into account environmental costs and other "externalities" associated with fossil fuels can tip the economic balance in favor of renewable energy sources in many more applications (Nogee et al. 1999).

Wind power came of age during the 1990s due to both technical advances and cost reductions. As costs declined and performance improved, grid-connected wind power capacity increased from about 2 gigawatts (GW) in 1990 to an estimated 25 GW by the end of 2001 (BTM Consult 2001). Wind power output worldwide is now growing about 30 percent per year. Furthermore, wind technology is continuing to improve and its costs are continuing to decline through the development of larger turbines, improved rotor blades and controls,

TABLE 1-3

Status of Renewable Energy Technologies

Technology	Rate of Expansion (%/yr)	Total Capacity at End of 1998	Energy Production in 1998	Investment Cost ($/kW)	Energy Cost in 1998	Potential Future Energy Cost
Biomass energy						
Electricity	~ 3	40 GWe	160 TWh	900–3,000	5–15 ¢/kWh	4–10 ¢/kWh
Heat	~ 3	> 200 GWth	700 TWh	250–750	1–5 ¢/kWh	1–5 ¢/kWh
Ethanol	~ 3	18 billion liters	420 PJ		8–25 $/GJ	6–10 $/GJ
Wind power	~ 30	10 GWe	18 TWh	1,100–1,700	5–13 ¢/kWh	3–10 ¢/kWh
Solar PV power	~ 30	0.5 GWe	0.5 TWh	5,000–10,000	25–125 ¢/kWh	5–25 ¢/kWh
Solar thermal electricity	~ 5	0.4 GWe	1 TWh	3,000–4,000	12–18 ¢/kWh	4–10 ¢/kWh
Low-temp. solar heat	~ 8	18 GWth	14 TWh	500–1,700	3–20 ¢/kWh	2–10 ¢/kWh
Hydropower						
Large	~ 2	640 GWe	2,500 TWh	1,000–3,500	2–8 ¢/kWh	2–8 ¢/kWh
Small	~ 3	23 GWe	90 TWh	1,200–3,000	4–10 ¢/kWh	3–10 ¢/kWh
Geothermal						
Electricity	~ 4	8 GWe	46 TWh	800–3,000	2–10 ¢/kWh	1–8 ¢/kWh
Heat	~ 6	11 GWth	40 TWh	200–2,000	0.5–5 ¢/kWh	0.5–5 ¢/kWh

NOTES: GJ = gigajoule; GWe = GW of electric capacity; GWth = GW of thermal capacity; kWh = kilowatt-hour; PV = photovoltaic; TWh = terawatt-hour.
SOURCE: Turkenburg 2000.

offshore applications, and other innovations. Wind-generated electricity now costs $0.04 to 0.06/kWh, depending on wind speed, and could eventually cost as little as $0.02 to 0.03/kWh (McGowan and Connors 2000, Short 2002).

Researchers and private companies have greatly improved the performance and reduced the cost of solar PV modules over the past 20 years. The cost of complete PV systems was on the order of $5 to $10 per watt as of 2000, leading to PV-generated electricity costing from $0.25/kWh to $1.25/kWh (Rever 2001). These costs, while high relative to typical electricity costs in industrialized nations, make PV power economical in certain niche applications such as off-grid rural areas where it is costly to extend transmission and distribution lines. Also, grid-connected PV applications are rapidly expanding in a number of countries including Germany and Japan.

The global PV market nearly quadrupled from 1995 to 2000 due to a combination of technical improvements and supportive policies (Maycock 2001). To maintain high growth, PV companies are developing newer types of PV cells and new products such as building-integrated PV modules. Given the opportunities for further technological improvement and market expansion, the cost of PV systems could fall to $2 to $5 per watt between 2005 and 2010 and around $1 to $3 per watt in the 2015 to 2020 time frame (Turkenburg 2000). If these cost targets are met, PV power would become economical in a much wider range of applications.

Biomass energy includes wood, forestry and agricultural residues, and crops grown for energy use. Using biomass for energy is carbon neutral as long as the biomass is produced on a sustainable basis. Plants and trees absorb carbon dioxide from the atmosphere through photosynthesis, although additional energy is required for biomass production and processing. Biomass can be burned for space heating, cooking, or production of electricity and/or useful thermal energy. Biomass can also be converted to gaseous and liquid fuels, as is done in Brazil, China, India, and the United States (see Chapter 4 for the Brazil case study).

Many countries generate electricity and useful thermal energy using residues from the paper and wood products industries as well as from municipal solid wastes. In the United States, electricity generation from bioenergy doubled between 1987 and 1999 (Short 2002). A few countries such as Sweden have begun to produce dedicated energy crops such as willow tree plantations. Sweden in fact obtained 17 percent of its total energy supply from biomass as of 1999 and intends to increase this share to 40 percent by 2020 (Turkenburg 2000).

New technologies for producing and processing biomass are under development including new types of energy crops, optimized production systems, gasification technologies, and new methods for converting cellulosic materials to ethanol. Also, researchers and private companies are developing methods for converting biomass to valuable chemicals as well as fuels such as hydrogen. These technologies are relatively expensive today, but continued R&D should reduce their cost and expand the market for biomass energy in the future (Kheshgi, Prince, and Marland 2000).

Geothermal energy uses the heat within the earth to produce steam, hot water, and power. About 45 billion kWh of electricity and nearly as much useful thermal energy were produced worldwide from geothermal resources as of 1998 (Turkenburg 2000). Geothermal plants produce a large fraction of the energy consumed in Iceland, over 20 percent of the electricity generated in the Philippines, and 5 percent of the electricity used in California. Geothermal energy production worldwide is expanding about 5 percent per year, with this growth rate limited by the scarcity of high-quality geothermal resources. Developing advanced technologies such as utilization of low temperature resources or drilling deep wells and pumping down water to extract the heat from hot underground rocks could accelerate the growth of geothermal energy (Mock, Tester, and Wright 1997).

Hydroelectric plants provide about 2.6 trillion kWhs per year, about one-fifth of all electricity produced worldwide. But much of the cost-effective, acceptable hydropower potential is already exploited in OECD nations. Developing countries, on the other hand, still have considerable untapped hydropower resources. Less than 7 percent of the hydropower potential in Africa has been harnessed, for example (Karekezi 2002a). Considering this untapped potential, it is estimated that hydroelectric production could rise to 6 trillion kWhs per year by the middle of this century (Turkenburg 2000). But environmental and social pressures (e.g., opposition to hydro projects that would displace large numbers of people) could limit hydropower growth.[4] At the same time, new technologies are being developed to minimize the adverse social and environmental impacts of hydropower (Marsh and Fisher 1999, Turkenburg 2000).

Energy Efficiency Opportunities

Energy efficiency improvement—using less energy for a given task—is an important energy "resource" worldwide. A vast number of energy efficiency improvements in appliances, lighting devices, vehicles, buildings, power plants, and industrial processes were developed and introduced over the past few decades. Adoption of these technologies is growing, contributing to a substantial reduction in energy intensity in many countries.

Figure 1-6 shows the progress made in reducing overall energy intensity—

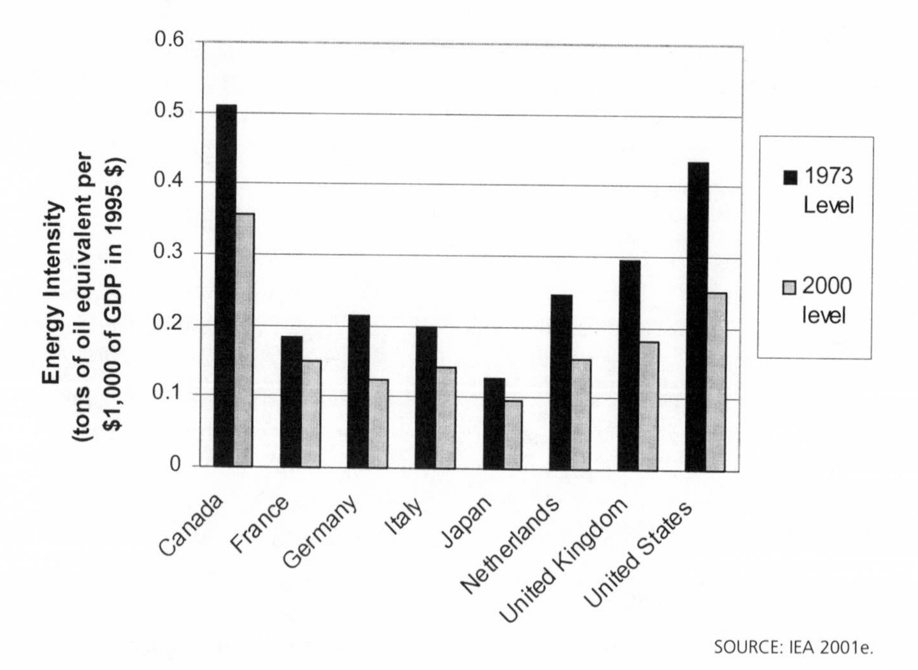

SOURCE: IEA 2001e.

FIGURE 1-6: Energy intensity reductions in industrialized nations.

energy use per unit of economic output (GDP)—in eight major OECD countries since 1973. Over this period, energy intensity declined 43 percent in Germany, 42 percent in the United States, 39 percent in the United Kingdom, and 24 percent in Japan. Structural changes such as the shift in economic output from heavy industries to light industries and the services sector caused some of these reductions, but much of the decline was due to real energy efficiency gains (IEA 1997d, Schipper et al. 2001).

Far greater energy savings could be achieved through widespread adoption of commercially available and cost-effective energy-efficient technologies such as:

- More efficient cooking and water heating devices;
- Buildings that make better use of natural lighting and ventilation;
- Building that have reflective roofs to reduce their air conditioning load;
- Compact fluorescent lamps replacing incandescent lamps;
- Other energy-efficient lighting devices;
- Refrigeration equipment with more efficient compressors and better heat exchangers;
- Energy management and control systems;

- Electronic devices with low standby power consumption;
- Improved design and control of pumping, compressed air, and other motor systems; and
- More efficient vehicle engines and drivetrains, lighter-weight vehicles, and hybrid electric vehicles.

These are just some of the technologies that can lower the amount of energy consumed for a given task. There are many other options for reducing energy use in applications ranging from washing clothes to operating electronic equipment to manufacturing steel, chemicals, and other basic materials. In addition, more intelligent design of new homes, office buildings, vehicles, and industrial processes through an integrated "systems engineering approach" can provide substantial energy savings, often with first cost savings as well (Hawken, Lovins, and Lovins 1999).

There is large potential for saving energy at lower cost than supplying energy in every nation. Table 1-4 shows the cost-effective national energy savings potential in the United States based on major studies conducted by leading national laboratories. The energy savings potential is 33 percent or greater in the case of residential lighting and refrigeration, heating and cooling in commercial buildings, and new cars and light trucks, and at least 20 percent in various other applications. Studies for western Europe and Japan indicate similar cost-effective energy savings potential (Jochem 2000). It should be recognized, however, that the energy savings cannot be implemented overnight. It will take many years for the capital stock to turn over in order to achieve national savings of this magnitude.

The savings potentials shown in Table 1-4 are based on commercially available technologies as of the late 1990s. But the potential for energy efficiency improvement is not static. New energy-efficient appliances, lighting products, building components, and vehicles are routinely developed and commercialized around the world (Martin et al. 2000, Nadel et al. 1998). Companies and research institutes are also developing more efficient and cleaner ways of producing steel, aluminum, paper, chemicals, and other basic materials, as well as superconducting materials that could greatly reduce energy losses in electricity transmission and distribution. Furthermore, the materials and the digital revolutions offer the possibility for substituting high-tech, lightweight materials or the flow of information for energy-intensive materials and activities (Hawken, Lovins, and Lovins 1999). In addition, materials recycling and product reuse are on the rise. All of these innovations mean that the energy efficiency "resource base" is constantly expanding.

The potential for large energy savings is not limited to industrialized

TABLE 1-4

Cost-Effective Energy Savings Potential in the United States

Sector and End-Use	Energy Savings Potential (%)
Residential[a]	
Lighting	53
Refrigeration	33
Water heating	23–28
Space heating, new homes	19–39
Space heating, existing homes	11–25
Space cooling	16–23
Commercial[a]	
Space heating	48
Space cooling	48
Refrigeration	31
Lighting	25
Water heating	10–20
Transport[a]	
New cars	35
New light trucks	33
Industrial[b]	
Iron and steel	32
Aluminum	30
Paper and pulp	25
Glass	24
Cement	21

NOTES: [a]Average savings from best available, cost-effective technologies relative to typical new devices. [b]Average savings possible by 2010 from accelerated adoption of available, cost-effective technologies relative to maintaining 1997 energy intensity levels.
SOURCES: Interlaboratory Working Group 1997, Interlaboratory Working Group 2000.

countries. A few developing countries, most notably China and some southeast Asian nations, have cut their energy intensity in recent decades. Nevertheless, energy waste is widespread, and thus the potential to increase energy efficiency is enormous in developing and former communist nations (PCAST 1999). Homes in these countries typically have inefficient cooking devices and heating systems. Businesses usually contain inefficient lighting, motors, and industrial processes. Also, inefficient power plants and electricity distribution systems are common. The overall cost-effective energy savings potential in developing and transition nations is estimated to be 40 percent or greater (Jochem 2000).

Energy efficiency improvement is possible in energy supply and conversion as well as end use. New combined-cycle power plants provide electricity conversion efficiencies of 50 to 55 percent, approaching twice the level of older thermal power plants.[5] Combined heat and power systems (also known as cogeneration systems) can convert 80 to 90 percent of the fuel input into useful energy (Casten 1998). And fuel cell technologies under intensive development worldwide promise low emissions as well as high conversion efficiency, especially if designed to cogenerate electricity and useful thermal energy.

The Role for Natural Gas

Natural gas is the cleanest and least carbon-intensive fossil fuel. Where gas supplies are available, natural gas–fired, combined-cycle power plants are often the preferred option for expanding electricity generation. Combined-cycle, gas-fired plants are significantly more efficient and have lower capital costs than new coal-fired plants. Gas-fired power plants do not emit SO_2 or particulates, and their nitrogen oxide (NO_x) emissions are about 90 percent lower than those of new coal-fired power plants per unit of electricity output. In addition, their carbon dioxide emissions per kWh are 55 to 65 percent lower than those of coal-fired power plants (Williams 2000).

Natural gas reserves are plentiful and more widely distributed than oil reserves. In fact the transition countries of the former Soviet Union had slightly more proven gas reserves than Middle Eastern countries at the end of 2000 (BP 2001). Proven natural gas reserves worldwide more than doubled over the past 25 years due to the growing interest in natural gas, discovery of new gas fields, and improved exploration and production techniques. Given its advantages relative to other fossil fuels, natural gas use is increasing worldwide and is expected to continue to rise for decades. Natural gas is projected to increase from 20 percent of global energy supply in 1997 to 24 percent by 2020 in the IEA *Reference Scenario* (IEA 2000a). Natural gas provides nearly 28 percent of global energy supply by 2050 in the Low Growth, Low Carbon Scenario developed by IIASA-WEC (see Fig. 1-5).

The consumption of natural gas declines in absolute and percentage terms in the second half of the twenty-first century in sustainable future scenarios such as the IIASA-WEC Low Growth, Low Carbon Scenario. Solar energy, biomass, and other renewable sources eventually replace natural gas and other fossil fuels to become the dominant forms of energy in these scenarios. Thus natural gas is viewed as a "bridging fuel" to help the world make a smooth transition to renewable energy sources. While it is important to reduce petroleum and coal use starting now, natural gas use could rise for a number of decades in a future where regional and local air pollution is significantly cut, the risk of climate change is reduced to tolerable levels, global security is enhanced, and modern energy services are extended to all citizens of the world.

Natural gas is not always located near centers of demand and is costly to transport long distances. The high cost of building natural gas pipelines and other gas supply infrastructure will be especially challenging for developing and former communist nations. This means it is critical to use gas efficiently, whether in power generation or direct consumption in households, industries, and other sectors. Technologies for high-efficiency natural gas use include advanced combustion turbines, innovative cogeneration systems, fuel cells, gas-fired heat pumps, and highly efficient building designs (Nadel et al. 1998, Williams 2000).

What about Nuclear Power?

The 435 nuclear power plants operating worldwide provide about 17 percent of the global electricity output (IEA 2000a). But growth in nuclear power capacity has stalled for a number of reasons including lack of public support, the failure to develop long-term options for the safe disposal of nuclear waste, safety issues provoked in part by the Chernobyl nuclear accident, the potential for nuclear energy to contribute to nuclear weapons proliferation, and lack of cost competitiveness. New nuclear power plants have become uneconomical in part because of the development of lower-cost alternative technologies such as gas-fired, combined-cycle power plants and wind power.

Nuclear power does have the advantage of not emitting carbon dioxide or other greenhouse gases.[6] As the world moves forward with emissions limits to reduce global warming, there is some renewed interest in nuclear power. Nuclear power also has the advantage of not emitting air pollutants such as SO_2, NOx, and fine particulates that harm public health. However, it is widely acknowledged that nuclear power will not make a "comeback" and that nuclear power output will start to decline as older reactors are retired, unless all of the concerns listed here are adequately addressed and public support is regained (Beck 2001, PCAST 1999, Williams 2000).

Because of this realization, the nuclear energy industry is conducting R&D on new reactor designs and fuel cycles that have improved safety, lower cost, and less potential for contributing to weapons proliferation than current designs. These options include advanced light water reactors and gas-cooled reactors such as the pebble bed modular reactor (Williams 2000). Even if the technical goals of these efforts are realized, these reactors would still generate spent fuel and high-level nuclear waste. And there is no guarantee that the public would accept a new generation of nuclear power plants, even if specialists are satisfied that the technical problems are solved. New concerns regarding the vulnerability of nuclear reactors and waste storage sites to a terrorist attack present yet another challenge to the revival of nuclear power. Finally, even if the technical problems are resolved and public support is obtained, there is no guarantee that a new generation of nuclear power plants will be economical compared to alternative sources of electricity—sources that are advancing and declining in cost.

Given these wide-ranging and substantial challenges, a nuclear power revival does not look promising. Consequently, governments must decide whether to continue supporting nuclear energy R&D with the hope of "keeping the nuclear energy option open." The nuclear energy industry has received vast amounts of government support, both for R&D and for other subsidies, over the past 50 years. In light of nuclear energy's problems and the emergence of other more promising energy options, government support for nuclear energy R&D has significantly declined in nearly all nations (IEA 2001e).[7]

Economic and Social Context

Energy is more than power plants, fuels production, energy distribution systems, and end-use technologies. Energy has social and economic dimensions, including the institutions that produce, market, and service energy technologies. Energy also fits into a broader social context. It is important to recognize the major economic and social trends occurring worldwide, such as increasing globalization, energy sector restructuring and privatization, rapid technological innovation, and urbanization. All of these trends will influence energy strategies and systems in the twenty-first century, and all should be taken into account as policies for an energy revolution are crafted and implemented.

Increasing Globalization

Although globalization is not benefiting all nations and regions (particularly impoverished ones), it is a fact that trade barriers are falling and world trade is growing. The share of traded goods and services in total world economic output was about 43 percent in 1996, up from 25 percent in 1960 (Rogner and Popescu 2000). Energy trade approached 55 percent of global primary energy use in the late 1990s. Most of this is petroleum, but increasing amounts of coal and natural gas are traded among nations.

The global economy is becoming more integrated through mergers and acquisitions, joint ventures, the telecommunications revolution, and the expansion of multinational companies. Multinational companies are playing an increasing role in fossil fuel production and distribution, ownership of gas and electric utilities, and manufacturing of renewable energy and end-use technologies. As companies and markets become increasingly international, so too should policy interventions through coordinated action and policy harmonization.

Restructuring and Privatization

Many nations are privatizing formerly government-owned utilities, petroleum and natural gas companies, and other institutions. This is being done ostensibly to reduce inefficiency and attract private capital to the energy sector. At the same time, efforts are being made to increase the transparency of policymaking, reduce government subsidies, and open up markets to greater competition. Successful policies, whether energy sector restructuring, incentives for energy efficiency and renewable sources, or regulations, must engage the private sector and catalyze private investment in the desired technologies on a large scale.

Restructuring and privatization do not guarantee that investments in new energy supplies will occur, that efficiency and energy services will be improved, or that costs will be reduced. This point was illustrated by the electric sector debacle in California in 2000–01 when electricity prices soared and rolling blackouts occurred over a number of months due in part to anticompetitive practices by owners of newly privatized power plants (Cavanagh 2001). Power shortages and/or cost increases have also occurred in developing countries such as Brazil and India following electric sector restructuring and privatization (Reddy 2001). And with a few exceptions (e.g., Ghana and South Africa), power-sector reform has done little to expand electrification among poorer households in Africa (Karekezi and Kimani 2002).

Rapid Technological Innovation

Technological innovation is accelerating and affecting living standards, commerce, and communications around the world. The microelectronics revolution influences energy use and energy policy in a number of ways. One way is by increasing access to information through dissemination of personal computers, the Internet, and other telecommunications advances. The growth of information technologies in turn affects energy use both directly through improved appliance and process controls and through the use of electricity by electronic devices, and indirectly through structural changes that save energy (Romm, Rosenfeld, and Herrmann 1999).

Rapid technological innovation is also occurring in energy production, conversion, and end-use, including fossil fuel–based electricity generation, renewable energy production and conversion, oil and gas production, vehicle and appliance design, and manufacturing processes. However, the scope of these innovations is not equally shared, with a growing "technology gap" between rich and poor nations as well as between richer and poorer segments of some nations (Reddy 2000).

Urbanization

The fraction of the world's population living in urban areas was about 46 percent as of 1996 and is expected to rise to 61 percent by 2030. Most of the population growth projected over the coming decades will occur in the world's cities, and about 90 percent of this growth will occur in developing countries. It is expected that there will be 23 metropolitan areas with a population of at least 10 million by 2015. Most of these "megacities" will be located in Asia (UNFPA 2001).

The proliferation of giant cities like Tokyo, Mexico City, Sao Paulo, Mumbai, and Shanghai presents special challenges related to transport, air quality, provision of basic services, and employment. As urbanization accelerates, energy use in urban areas rapidly increases. On the other hand, urbanization can facilitate the dissemination of new energy technologies and presents the opportunity to design more efficient and sustainable habitats and transport systems as cities expand.

Summary

A business-as-usual, inefficient, and fossil fuel–intensive energy future is not desirable. It will cost too much, worsen air pollution, cause dangerous climate change, rapidly deplete precious oil resources, increase security risks, and exacerbate tensions among nations. It is unjust for 20 percent of the world's population to increase its already high use of fossil fuels while one-third of the world's population gathers twigs, stalks, and dried animal manure to cook meals and stay warm. These are serious problems today, and they will grow worse in the future and put the well-being of future generations at risk unless a different path is pursued.

A different path is possible, one that emphasizes higher energy efficiency, much greater reliance on renewable energy sources, and meeting the energy needs of the poor. This energy path would shift the world from fossil fuels to renewable energy sources over a number of decades, and would make greater use of natural gas during the transition.

The remainder of this book discusses the obstacles to and strategies for a more sustainable energy future. Chapter 2 reviews the barriers to greater energy efficiency and renewable energy use worldwide. Chapter 3 presents the policy options that can overcome these barriers. It includes numerous examples of implementation of these policies in different countries, and the lessons these experiences provide. Chapter 4 presents 10 case studies of greater energy efficiency or renewable energy use on a large scale in different parts of the world, in both industrialized and developing nations. All the case studies illustrate market transformation through a comprehensive set of policy and program initiatives.

Chapters 5 and 6 switch to a prospective focus. Chapter 5 presents and analyzes 10 major policies that could put the United States on a path toward a more sustainable energy future. Chapter 6 presents and analyzes a similar set of energy policies for a key developing nation—Brazil. Chapter 7 examines international policies and institutions for increasing energy efficiency and renewable energy use. It also recommends ways to improve international efforts. The final chapter reviews the overarching lessons from these policy assessments and some of the global trends in energy efficiency and renewable energy use. It also presents a scenario for how a clean energy revolution might unfold during the twenty-first century and discusses other factors that could significantly influence this revolution such as population growth and lifestyle choices.

Notes

1. Renewable energy sources include biomass, solar energy, wind power, hydropower, and geothermal energy.
2. These energy scenarios were produced by a multinational team under the auspices of the International Institute for Applied Systems Analysis (IIASA) and the World Energy Council (WEC) (Nakicenovic, Grubler, and McDonald 1998).
3. Oil import dependence is the fraction of total oil consumption provided by net imports (imports minus exports).
4. The Three Gorges Dam in China is perhaps the most controversial hydroelectric project now under development. China is constructing this massive proj-

ect in spite of the fact that it will displace over one million people and have wide-ranging adverse environmental impacts in its region (Fackler 2002).

5. Efficiency based on the higher heating value of the fuel input.
6. Greenhouse gas emissions will occur, however, due to the energy consumed in nuclear fuel processing, nuclear power plant construction, waste disposal, and plant decommissioning. The same is true for building and producing the materials that go into renewable energy facilities. But for both nuclear power and renewable energy sources, the greenhouse gas emissions per kWh are reduced by 90 percent or more compared to coal-fired power plants and are also well below those of gas-fired power plants (Holdren and Smith 2000).
7. From 1989 to 2000, the U.S. Department of Energy reduced its funding for nuclear fission and fusion R&D by over 70 percent.

Barriers

If energy efficiency measures are cost-effective, why aren't they routinely adopted by consumers and businesses? Likewise, if renewable energy sources are much cleaner and more socially acceptable than fossil fuel and nuclear energy sources, why aren't they chosen more frequently when new energy supplies are needed?

A wide range of barriers limit the introduction and deployment of energy efficiency and renewable energy technologies throughout the world (Table 2-1). The significance of the different barriers varies among sectors, institutions, and regions. Some of the barriers will shrink as energy efficiency and renewable energy technologies advance and gain market share, but others are likely to persist unless directly confronted through policy interventions. Taken as a whole, these barriers are inhibiting the transition to a more sustainable energy future.

The discussion of barriers is organized in terms of barriers to greater energy efficiency and barriers to renewable energy use. Some of the specific barriers are common to both areas, but many affect one but not the other. It is important to understand the nature and scope of these barriers before considering the policies and programs for removing or overcoming them.

TABLE 2-1

Barriers to the Adoption of Energy Efficiency and Renewable Energy Technologies
• Limited supply infrastructure
• Quality problems
• Insufficient information and training
• Misplaced incentives
• Lack of money or financing
• Purchasing procedures
• Pricing and tax barriers
• Regulatory and utility barriers
• Political obstacles

Barriers to Greater Energy Efficiency[1]

Limited Supply Infrastructure

Energy-efficient technologies are not produced or readily available in some countries, especially developing and former communist nations. Likewise, these countries often lack energy service companies that specialize in energy efficiency projects and investments. This creates a vicious circle where, because demand is low, suppliers do not make products or services available, and demand remains low due to limited availability.

Energy-efficient devices such as high-efficiency lighting products, appliances, motors, and control systems are difficult to find or nonexistent in many developing countries. For example, high-efficiency motors, motor-driven devices, and electronic adjustable speed drives are not readily available in China (Nadel et al. 2001). In some cases, energy-efficient products are produced in developing countries for export but not offered in the local market. For example, manufacturers produce or assemble high-efficiency air conditioners and motors in Brazil, but either they do not sell these models locally or they market them there on a very limited basis (Geller 2000).

In many cases, energy efficiency is combined with other features of an appliance, building, vehicle, or other product. Consumers are not offered the choice of an energy efficiency upgrade even if they are willing to pay more for this option. This is often the case for cars and appliances such as refrigerators and freezers. If higher energy efficiency is not a discrete option, consumers tend to ignore energy performance, and manufacturers have little incentive to improve efficiency.

Quality Problems

Some energy efficiency measures do not live up to performance claims due to manufacturing flaws, installation problems, or improper use. Many air conditioning systems in the United States, for example, are oversized or installed improperly, increasing their energy consumption (Neme, Proctor, and Nadel 1999). And the energy efficiency of many U.S. buildings is often below design expectations due to energy efficiency measures not being well integrated, properly installed, or commissioned at all (Aitken 1998). If these problems exist in the United States, they most likely exist in many other countries.

The poor quality of energy-efficient products is a problem in a number of developing countries. In China, for example, some energy-efficient lamps and lighting ballasts produce low light output, experience premature failure, and suffer from other problems This hurts the reputation of energy-efficient products in general and results in low consumer acceptance (Nadel et al. 1999).

The poor quality of energy audits and other energy efficiency services has been noted in a number of countries. For example, the quality of home energy audits and improper installation of efficiency measures has been identified as a problem in the United Kingdom (Crowley 2001). Also, the poor quality of industrial energy audits has been noted as a barrier to energy efficiency improvement in Thailand (Vongsoasup et al. 2002).

Insufficient Information and Training

Consumers and businesses may not be aware of energy efficiency options. And if they are aware of these measures, they may not have information on how much energy and money would be saved, or the extent of other benefits such as health benefits or productivity improvements. Consumers may be skeptical of energy savings claims made by vendors, contractors, or energy service companies, or they may have doubts whether an energy savings device will work properly in their home, commercial building, or factory.

Consumers pay a monthly utility bill and consequently do not know how much it costs to run each type of appliance. Likewise, consumers generally are not able to determine how quickly a particular energy efficiency measure, such as an energy-efficient appliance, furnace, or water heater, will pay back its incremental first cost. And future energy prices, which affect the economic viability of an energy efficiency project, are uncertain.

In the commercial and industrial sector, plant engineers and facility managers may not understand how to optimize the energy efficiency of industrial processes or buildings systems. Many firms, especially small and medium-sized enterprises, suffer from a shortage of trained technical staff. Architects and builders may lack the know-how to design and build state-of-the-art or even good-practice energy-efficient buildings. And contractors may not know how to properly size or install key technologies such as high-efficiency air conditioning or pumping systems.

Misplaced Incentives

The financial interests of those responsible for purchasing energy efficiency measures may not be aligned with those who would benefit from the purchase. In rental property, for example, the owner is usually responsible for equipment purchases, whereas tenants pay the energy bills. The incentive for the owner is to minimize up-front costs, leading to the purchase of inefficient equipment and a lack of energy efficiency upgrades. Indeed, energy consumption and expenditures per unit of floor area are much greater in rented apartment buildings and in public housing compared to owner-occupied, single-family housing in the United States (DeCicco et al. 1995).

In new construction, builders and contractors are usually hired on the basis of the lowest-priced bid. New homes and commercial buildings are rarely designed to minimize lifetime cost, taking into account energy and other operating costs as well as first cost. Unless building energy codes require efficiency measures, they are usually not specified by the designer. Also, builders and contractors often "cut corners" to save money, such as improperly installing insulation or heating and air conditioning duct work. And public officials may lack the resources or motivation to enforce building energy codes. This combination of perverse incentives rarely leads to the design and construction of energy-efficient, well-integrated buildings (Lovins and Lovins 1997).

Misplaced incentives can take other forms. In Russia, many residential and commercial buildings lack heat or gas meters, and residents pay a fixed monthly bill for heating energy. Consequently, there is no incentive to conserve energy nor any means of measuring savings (Martinot 1998). Likewise, some nations, states, or municipalities receive a significant share of their budgets from taxes on energy production and/or consumption (Jochem 2000). This diminishes their enthusiasm for promoting energy efficiency.

Purchasing Procedures

Many buildings are constructed, products purchased, or facilities renovated on the basis of least first cost, not least life-cycle cost. Consumers tend to consider the lowest-cost product with the desired features the "best buy." These behaviors discourage inclusion of energy efficiency measures, even if they would provide a rapid payback.

Energy efficiency is highly decentralized and pervasive. Millions of consumers and businesses make decisions concerning energy efficiency when they buy appliances, lights, or vehicles; construct new buildings; or expand manufacturing capacity. Energy efficiency is often ignored or given low priority when these decisions are made. Research in the United States, for example, shows that consumers pay more attention to price, product features and performance, capacity, reliability, and brand name than energy efficiency when selecting an appliance (Shorey and Eckman 2000). Also, some purchases are made to replace a failed product—for example when a heating system fails during the winter, an air conditioner fails in the summer, or a motor fails in an industrial plant. These purchases are often made in a hurry, with no time allowed to search for a high-efficiency model.

Energy costs often represent a very small fraction of the total cost of owning and operating a factory or business. In the United States, for example, energy costs represent 1 to 2 percent of the total cost of production for manufacturing industries other than energy-intensive industries such as aluminum, steel, and paper production (Geller and Elliott 1994, Lovins and Lovins 1997). Businesses are most concerned with developing new products, maintaining production, and increasing sales; energy consumption is a secondary or tertiary concern. And many firms are risk averse and perceive energy-efficient technologies as risky.

These factors lead firms to establish internal hurdle rates for energy efficiency investments that are much higher than their cost of capital or the hurdle rates used for other types of projects[2] (DeCanio 1993). In other words, many companies do not take advantage of opportunities to raise their profits by increasing energy productivity. While this may not present problems for individual businesses, inefficiency and excess energy consumption present problems for society as a whole.

Lack of Money or Financing

Consumers or businesses may lack the money needed for an energy efficiency project, and financing may not be available. Longer-term commercial lending is rare or nonexistent in many developing and former communist nations (Martinot 1998). Likewise, energy service companies may not be able to provide or arrange financing for potential clients, especially in developing countries.

Even in industrialized countries, financing at attractive interest rates may not be available for energy efficiency measures or projects, even if they are modest in scale. Consumers or businesses that are heavily in debt may be unable or unwilling to pay the extra first cost for a more efficient product. Fossil fuel producers and electric utilities, on the other hand, have access to long-term financing at low interest rates for new power plants or other energy supply projects.

Pricing and Tax Barriers

Many nations, both rich and poor, subsidize conventional energy sources. The U.S. federal government provided about $145 billion (in 1999 dollars) in subsidies to nuclear power from 1947 to 1999 (Goldberg 2000). The U.S. government also provided tax incentives worth about $140 billion to oil companies from 1968 to 2000 (GAO 2000).

Subsidies for the fossil fuel and electric industries have been scaled back but continue to be provided in many countries. Oil companies still received about $1.5 billion in tax incentives in the United States in 2000 (GAO 2000). Germany provided about $5 billion per year in subsidies to its coal producers as of 1998 (IEA 2000f). India provides over $3 billion in subsidies to electricity consumers, mainly agricultural consumers (Shukla et al. 1999). Agricultural consumers in India typically paid only 0.5 cents per kWh as of 1998, about one-tenth the average cost of electricity supply (Jochem 2000). Subsidized energy prices reduce the economic incentive for consumers to use energy efficiently.

Many former communist nations still heavily subsidize energy, although these subsidies were reduced in recent years. Russian municipalities, for example, spent 25 to 45 percent of their budget on residential heat subsidies in the late 1990s, covering more than half of heat bills (Jochem 2000). In addition, a significant fraction of households and businesses in former communist nations do not pay their energy bills.

Even if energy prices are not subsidized, they rarely reflect the full costs to society associated with energy production and use, including social and environmental costs. Electricity prices do not include the full costs to society from air, water, and land pollution or global climate change caused by power generation and supply.[3] Prices for gasoline and other petroleum products do not include the costs to society from the pollution created when the fuel is produced and burned, the military expenditures to protect oil supplies, or the economic upheaval caused by periodic oil price shocks. The failure to include the cost of these social and environmental "externalities" in energy prices leads to excessive consumption, relative to what would be socially desirable.

Energy prices are generally based on average rather than marginal costs, and usually do not reflect the cost of producing energy at different times of the day or year. The cost of supplying electricity, for example, can be much higher during periods of peak demand compared to other times. But if consumers are not charged time-of-use rates, they do not have a financial incentive to save energy during peak periods or shift their electricity use to less costly periods.

Tax policies can also discourage investment in capital-intensive energy efficiency technologies. In the United States, for example, businesses can deduct energy costs from their revenues before determining their income tax. Capital costs, however, must be depreciated over many years, in some cases 30 years or more. This makes it more difficult for businesses to justify undertaking energy efficiency projects. Likewise, some states charge sales taxes on energy-saving devices but not on fuels and electricity.

Regulatory and Utility Barriers

In most cases, utilities increase their profits when they sell more electricity or natural gas, and lose money if consumers reduce their energy use. Utilities do not have a financial incentive to promote more efficient energy use, even if this is in the best interest of consumers and society as a whole. This barrier results from the way utility rates and profits are normally regulated (Hirst, Blank, and Moskovitz 1994).

Utility privatization and deregulation can also inhibit utility investment in efficiency measures. Utility investment in demand-side management activities declined nearly 50 percent in the United States between 1993 and 1998 as deregulation occurred or was considered in many states[4] (Nadel and Kushler 2000). Utilities cut their energy efficiency programs and other nonessential costs as competition in power generation heated up and regulation diminished.

Privatizing utilities and separating power generation, transmission, and distribution, as is occurring in a number of countries, can make this situation worse. Distribution utilities have even less financial incentive to invest in end-use efficiency than vertically integrated utilities because their profits will decline even more for a given reduction in electricity sales (Cowart 2001). Newly privatized utilities are usually concerned with producing short-run profits rather than maximizing longer-term social and economic benefits for society as a whole, given the profit orientation of investor-owned utilities or other businesses that are purchasing public utilities around the world (Clark 2000).

Utilities can also erect barriers that inhibit on-site cogeneration of electricity and thermal energy, an efficient energy strategy. Utilities can refuse to enter into long-term contracts at reasonable rates for buying excess power from cogeneration projects. Utilities can adopt excessive interconnection requirements or overprice backup power from the central grid. Also, utilities can charge onerous "exit fees" if businesses leave the centralized power grid to generate power on-site (Casten 1998).

Political Obstacles

Powerful industries can oppose and block political action to increase energy efficiency. For example, coal and oil producers along with energy-intensive industries oppose the adoption of fossil fuel or carbon dioxide taxes. Auto producers oppose vehicle efficiency standards or taxes on "gas guzzlers." Likewise, builders and appliance manufacturers tend to oppose minimum efficiency standards on their products. These business interests have a great deal of political clout and are highly motivated to block the adoption of policies that are perceived to be harmful. The potential "winners" from adopting these policies, on the other hand, are often less organized, less motivated, and less influential in the political process.

The failure to adopt stronger auto fuel economy standards in the United States in the past 15 years, in spite of strong public backing for this action, is a good example of this dilemma. Another example is the experience with energy or emissions taxes in the United States. The fossil fuel industries contribute heavily to political campaigns, and their political influence has blocked the adoption of higher energy taxes or taxes on carbon dioxide emissions.[5]

These barriers, taken as a whole, lead to what is sometimes termed the "efficiency gap" where consumers and businesses implicitly require a rate of return of 30 percent or greater (i.e., a payback of three years or less) before investing in energy efficiency measures. This rate of return is much higher than the social or market cost of capital. It should be remembered that this "efficiency gap" is inferred from actual purchasing decisions and behavior—decisions that reflect the host of market failures and obstacles described above. It does not reflect a conscious choice on the part of consumers or businesses.

Barriers to Renewable Energy Use[6]

Limited Supply Infrastructure

Small-scale renewable energy technologies such as solar heating and electricity systems may not be readily available, particularly in rural areas where they can be most economically viable. Also, larger scale wind and biomass conversion technologies may not be available in some countries. The demand for renewable energy technologies may be too low or too diffuse to justify local production, import, or marketing. This creates a vicious circle—private firms are reluctant to enter the renewable energy business in new regions where the technology is not yet established, and the market never gets established without the presence of equipment suppliers.

Although production of many renewable energy technologies is increasing, it is still not large enough to achieve significant economies of scale and rapidly drive down production costs in some cases. With limited production and sales, marketing and transaction costs can be high. And as long as prices are high, demand will remain limited. Solar photovoltaic (PV) modules, for example, still exhibit costs such that demand is relatively inelastic and limited to a variety of niche applications, although efforts to widely promote and install PV systems are expanding (Oliver and Jackson 1999).

Renewable energy technologies can be costly in countries where they are not yet manufactured, relative to locally produced energy sources. Grid-connected wind power has not made a major contribution so far in China, for example, because large-scale wind turbines are not yet produced in China, and the cost of electricity from imported turbines is relatively high (Lew et al. 1998, Lew and Logan 2001). Also, the absence of local PV module production increases the cost and limits the market for PV systems in many developing countries. Import duties on renewable energy technologies and components can contribute to this problem.

Quality Problems

Some renewable technologies such as solar PV and bioenergy systems lack standardization and quality control. For example, the growth of the PV market in Africa has been retarded by the poor quality of many products (Simm, Haq, and Widge 2000). Systems can be improperly assembled or installed, thereby degrading performance. Likewise, service and repair capabilities can be lacking or inadequate. For example, the poor installation, maintenance, and repair capability for rural PV systems has been noted as a serious problem in Kenya, Zimbabwe, and South Africa (Martinot et al. 2002, Simm, Haq, and Widge 2000).

Insufficient Information and Training

As is the case for energy efficiency measures, consumers may be unaware of renewable energy options, local product suppliers, or financing opportunities. Likewise, consumers may lack credible information on the performance, reliability, or economic merit of renewable energy options. And obtaining this information can take time and/or money.

On the supply side, renewable energy project developers need accurate data about wind, solar, bioenergy, and geothermal resources. This is critical for properly siting, sizing, and installing renewable energy systems. But renewable energy resource assessments are lacking in some regions. This has been noted as one of the reasons limiting the development of wind power in China, for example (Lew and Logan 2001).

Renewable energy businesses may lack information on potential customers and their willingness to adopt renewable technologies. This lack of market information can be especially problematic for companies trying to sell newer technologies such as solar PV or distributed bioenergy systems. In addition, utilities often lack information on how the output of renewable technologies such as wind and solar systems would affect their load, and in particular reduce peak load.

Lack of Money or Financing

Given that renewable energy technologies have a relatively long payback, it is critical to offer longer-term financing at low interest rates and with long loan terms. Traditional lenders such as national development banks or private banks have been reluctant to provide loans for renewable energy technologies because of small project size, unfamiliarity with the technologies, and other considerations. The access to credit is especially problematic in rural areas of developing countries where poorer households lack acceptable collateral.

Many low-income rural households already pay $3 to $15 per month for energy in the form of candles, kerosene, and batteries (Martinot, Cabraal, and Mathur 2000). They are willing to pay for higher-quality energy sources and services (e.g., solar PV systems or lanterns), but they need long-term credit at low interest rates to make this choice affordable.

The availability of low-interest financing can also make a major difference in the feasibility of renewable energy technologies in industrialized countries. In the United States, for example, financing of solar PV panels by publicly-owned utilities results in a cost for solar power that is nearly two-thirds less than the cost if the project is financed by private developers (Jones and Eto 1997).

Pricing and Tax Barriers

As is the case for energy efficiency measures, renewable energy measures are disadvantaged if the price for conventional energy sources is subsidized or structured so that it is not based on actual costs. In India, for example, it is very difficult for biogas-powered water pumps (or other renewable technologies) to compete with diesel and electric pumps when the price of rural electricity and diesel fuel is heavily subsidized (Martinot et al. 2002). Electricity prices often do not reflect the full cost of electricity grid extension in rural areas of developing countries. This discourages the adoption of decentralized renewable energy technologies such as solar PV systems, which in fact can be more cost effective than grid extension based on actual costs.

As noted above, energy prices rarely reflect the full costs to society associated with conventional energy production and use, including social and environmental costs. Likewise, buyback rates offered by utilities may not reflect all the benefits of renewables—for example, the value of supply diversification, increased system reliability, peak demand reduction, and so forth. These pricing distortions make it difficult for renewable energy sources to compete with conventional energy sources.

Tax policies can discourage the adoption of capital-intensive renewable technologies. This is the case when businesses are allowed to deduct fuel purchases from revenues when calculating income taxes, but must depreciate renewable energy devices over many years. Some countries subject imported renewable energy technologies or components such as PV cells and wind turbines to high import duties, thereby driving up their cost. And renewables can be discouraged by the tax breaks such as "depletion allowances" provided to conventional fossil energy resources.[7]

Regulatory and Utility Barriers

Utilities can impede renewable energy development by adopting onerous interconnection requirements, refusing to pay reasonable rates or provide long-term contracts for excess power provided to the electric grid, or establishing burdensome application procedures. Developers of small-scale renewable energy projects, say those 20 kW or less, do not have the time or money to negotiate terms for interconnection and power sales to the utility on a project-by-project basis.

Siting and approval of centralized renewable energy projects can also be time consuming and costly. For example, it is relatively easy to obtain approval for siting and installing large-scale wind turbines in Denmark and Germany, but quite difficult to do so in the Netherlands and Sweden (Jacobsson and Johnson 2000).

Political Obstacles

Many governments favor conventional fossil fuel sources and electric generation technologies over renewable energy technologies due to tradition, familiarity, and the size, economic strength, and political clout of the conventional energy industries. In the case of developing countries, other key institutions such as the World Bank and multilateral development banks have resisted lending for renewable energy projects due to their small size, complexity, higher perceived risk, and other factors (Martinot 2001).

As is the case for energy efficiency measures, vested interests can exert pressure in the political arena to block the adoption of policies favorable to renewable energy sources. Electric utilities, fossil fuel producers, and vendors of conventional energy technologies often oppose financial incentives or market reserves for renewables. This occurred, for example, when adoption of wind power expanded in Germany during the 1990s. Fortunately, wind turbine manufacturers and owners were able to defeat an effort by large utilities to weaken the financial incentives for wind power, although the vote in the Bundestag was very close (Jacobsson and Johnson 2000).

In the United States, most electric utilities oppose market reserves for renewable electricity and are preventing the adoption of such reserves at the national level and in many (but not all) states. Also, oil companies oppose and are preventing the adoption of market reserves for renewable-based fuels. Renewable energy industries are relatively immature and much less influential in the political arena than conventional energy suppliers.

This interrelated set of barriers can be especially problematic for renewable energy technologies since renewables have a difficult time competing with conventional energy sources in the marketplace today. Some of the barriers listed above inhibit off-grid applications, others apply more to grid-connected renewables. But without targeted policy initiatives to overcome these barriers, renewable energy sources in all likelihood will remain niche technologies that contribute relatively little to worldwide energy supply in the next few decades.

Summary

A host of barriers are limiting energy efficiency improvement and renewable energy adoption worldwide. Some are technical in nature (e.g., the limited availability of products or quality problems). Some relate to human behavior (e.g., the low priority given to energy issues or the tendency to purchase products based on least first cost). Other barriers are due to flaws in the ways markets operate (e.g., energy price subsidies, energy prices that do not include social and environmental costs, or poorly informed consumers). And others relate to public policies and institutions (e.g., the lack of attractive financing for efficiency and renewable energy measures, regulations that discourage energy efficiency or renewable energy use, or tax policies that disadvantage these technologies).

It is possible to remove many of the barriers discussed in this chapter through enlightened public policies—policies that eliminate price subsidies, make energy efficiency and renewable energy technologies readily available, improve the performance of these technologies, educate and train consumers, require certain levels of efficiency or renewable energy use, or provide convenient financing. But other barriers, such as the low priority given to energy issues or the tendency to focus on first cost rather than life cycle costs, cannot be easily removed through public policies. Nevertheless, policies such as financial incentives and regulations can overcome these pervasive barriers. The design and application of these public policies is the focus of the next two chapters.

Notes

1. In addition to the specific references listed in the text, this section draws from Brown 2001, Eto, Goldman and Nadel 1998, Jochem 2000, Martinot and McDoom 2000, and Reddy 1991.
2. The hurdle rate is the required rate of return on a potential investment.
3. Pollution control costs are embedded in energy prices, but much pollution is uncontrolled as is the emission of the "greenhouse gases" causing global climate change.
4. Demand-side management activities include energy efficiency and load shifting programs.
5. For example, the fossil fuel industries and their allies lobbied heavily to prevent the Congress from adopting a small energy tax proposed by President Clinton in 1993.
6. In addition to the specific references listed in the text, this section draws from IEA 1997b, Martinot and McDoom 2000, Nogee et al. 1999.
7. Depletion allowances are tax deductions that companies are provided based on extraction of fossil fuel reserves.

Policy Options

There is no "silver bullet" for overcoming the barriers to a more sustainable energy future. Many policy initiatives are needed to increase the availability and deployment of energy efficiency and renewable energy technologies. These policies can be grouped into the following 12 categories:

- Research, development, and demonstration
- Financing
- Financial incentives
- Pricing
- Voluntary agreements
- Regulations
- Information dissemination and training
- Procurement
- Market reforms
- Market obligations
- Capacity building
- Planning techniques.

Figure 3-1 illustrates the role of the different policies in relation to the market share of a particular technology (i.e., the classic S-shaped diffusion curve). Certain policies such as research and development, financial incentives, and procurement initiatives are most appropriate for stimulating commercialization and initial markets for new technologies. Other policies such as financing, voluntary agreements, and information dissemination are used to accelerate adoption once a technology is established in the marketplace. Policies such as regulations and market obligations are often used to maximize market share or complete the market transformation process. But there are many exceptions to these general rules (e.g., market obligations can be used to stimulate commercialization or incentives can be helpful throughout the process).

An integrated approach to market transformation often consists of a combination of "technology push" through research, development, and demonstration (RD&D); "demand pull" through financial incentives, education and training, procurement, or market obligations; and "market conversion" through codes and standards (Loiter and Norberg-Bohm 1999). In addition, some actions such as energy pricing reforms, capacity building, and planning techniques can facilitate the effective implementation of other more focused policies. An integrated

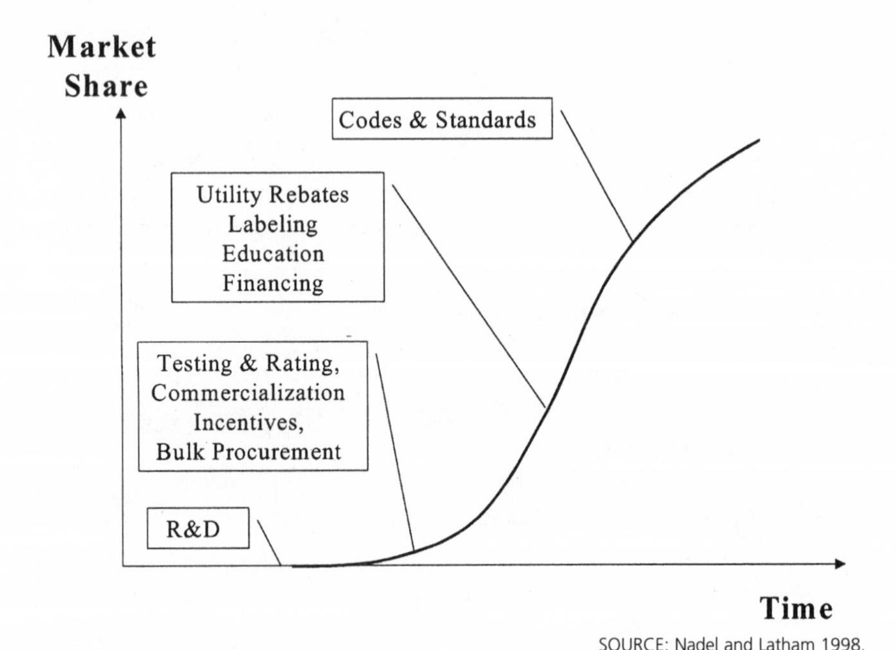

SOURCE: Nadel and Latham 1998.

FIGURE 3-1: Policy options to facilitate market transformation.

approach can also account for and address the multiple barriers that are likely to exist in any country or locale.

The appropriate mix of policies in any particular situation depends on technological attributes, the barriers that exist, and market conditions. This conceptual framework is sometimes referred to as the "innovation system" (Jacobsson and Johnson 2000). The innovation system consists of a wide range of factors including the knowledge base, the prices and relative performance of competing technologies, the behavior of different actors in the marketplace (such as technology suppliers and users), the networks among these actors (such as how new technologies move from suppliers to end users), institutions that can foster or impede innovation, and the cultural context.

In some cases it is possible to advance energy efficiency and renewable energy technologies through existing innovation systems. In other cases, the barriers are too great and it is necessary to create new systems (Jacobsson and Johnson 2000). These general concepts—the role of different policies in the market diffusion process, the need to integrate policies to overcome multiple barriers and transform markets, and the fact that policies are implemented within a wide-ranging and complex innovation system—are useful to keep in mind when considering the design of and experience with the specific policies discussed in the following text.

Research, Development, and Demonstration

RD&D is critical for expanding the knowledge base and maintaining a pipeline of new and improved energy supply and end-use technologies. Government-funded RD&D is justified on the basis of the private sector underinvesting in RD&D relative to what is considered desirable from a societal perspective. RD&D provides many public benefits that the private sector cannot capture, such as greater fuel diversity, reduced pollutant emissions, enhanced national security, and increased knowledge for society in general (NAS 2001a). Individual companies also tend to underinvest in RD&D because of their focus on short-term profits, concern that they will not be able to fully recover the costs for developing and commercializing new technologies, or fear that any competitive advantage will be short-lived (PCAST 1997).

Government-funded RD&D has helped to advance many energy efficiency and renewable technologies during the past 20 years. For example, government-funded RD&D contributed to innovations in wind turbines, lighting technologies, high-efficiency appliances, solar power systems, building technologies, advanced engines and turbines, and bio-based fuels in the United States (DOE

2000, Geller and McGaraghan 1998, Loiter and Norberg-Bohm 1999). The U.S. National Academy of Sciences estimates that 17 energy efficiency RD&D projects funded by the U.S. Department of Energy from 1978 to 2000 yielded about $30 billion in direct economic benefits, substantially exceeding the $7 billion in total federal funding for energy efficiency RD&D over this time period (NAS 2001a). These projects also provided substantial environmental, national security, and other indirect benefits.

The cost of wind power fell by a factor of 10, the cost of solar photovoltaic (PV) power fell by more than a factor of 10, and the cost of solar thermal power fell by more than a factor of 5 in the past 25 years, meeting or exceeding cost reduction projections (IEA 2000e, McVeigh et al. 1999). RD&D played a large role in these dramatic cost reductions. For example, it is estimated that RD&D produced over 70 percent of the cost reduction in PV modules in Japan from 1976 to 1990 (IEA 2000e). RD&D also helped to improve the performance and reduce the cost of wind power during the 1980s and '90s (IEA 2000e, Johnson and Jacobsson 2001, Loiter and Norberg-Bohm 1999).

Energy RD&D declined significantly in a number of industrialized countries during the 1980s and '90s (Dooley 1998; IEA 2000f). This decline is noticeable in both public and privately funded energy RD&D. In the United States, government support for energy technology RD&D declined 63 percent between 1980 and 1997 (Fig. 3-2). This trend is correlated with a decline in energy innovations as determined by patent registrations (Margolis and Kammen 1999). In addition, industrialized nations devoted the majority of their energy RD&D funding to nuclear energy and fossil fuels during the past 20 years. Relatively small shares were devoted to energy efficiency and renewable energy technologies.

Many studies call for greater public support for RD&D on clean energy technologies. President Clinton's Committee of Advisors on Science and Technology, for example, recommended doubling government-funded energy efficiency RD&D and estimated that this could produce a 40 to 1 return for the nation (PCAST 1997). The same committee also recommended increasing renewable energy RD&D by 240 percent given the potential impacts and benefits these technologies could have.

Energy RD&D priorities have shifted in a number of countries due to waning interest in nuclear power, growing concerns about global warming, and other factors. The fraction of government-sponsored energy RD&D devoted to energy efficiency and renewable energy sources rose in the United States, Japan, Germany, and other countries during the late 1990s (Dooley and Runci 1999, IEA 2000f). Japan, for example, increased its government support for RD&D of renewable energy, energy efficiency, and advanced coal technologies to about $400

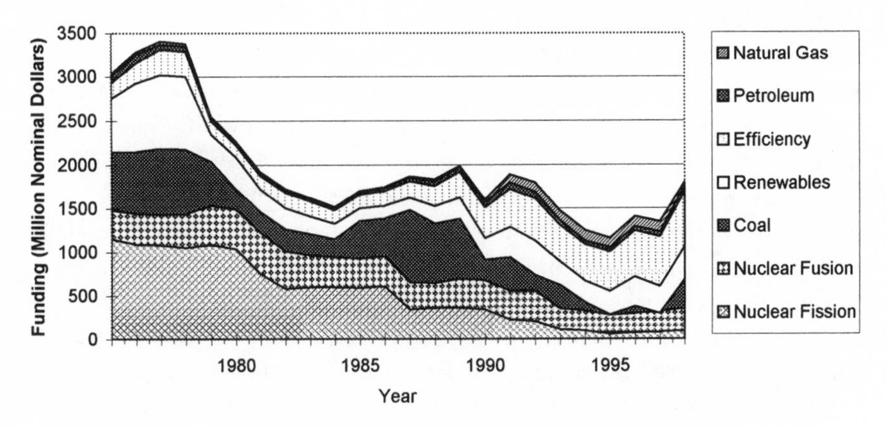

SOURCE: DOE 2001.

FIGURE 3-2: U.S. energy technology R&D funding, 1978–2001.

million per year as of 1998 (Katsumata 1999). In the United States, RD&D on energy efficiency and renewable energy increased to $1.0 billion or 56 percent of total government-funded energy RD&D in 2001, up from just 18 percent of the total in 1990 (see Fig. 3-2).

The experience with energy efficiency and renewable energy technology RD&D over the past 25 years provides a number of lessons. One is that RD&D can improve the performance and reduce the cost of innovative technologies, but RD&D alone may not result in their widespread adoption and use. However, RD&D in conjunction with "demand pull" policies such as financial incentives, market reserves, bulk procurement, or regulations can be a powerful force for technology innovation and deployment. The demand pull policies foster commercialization and market formation, which in turn can lead to further technological improvement, expanding markets, and cost reductions. This phenomenon is demonstrated by the experience with high-efficiency appliances, energy-efficient lamps and ballasts, wind power, and solar PV power in the United States, Western Europe, and Japan (Geller and McGaraghan 1998, NAS 2001a, Payne, Duke, and Williams 2001).

Another lesson is that collaboration between research institutions and the private sector can be an especially productive RD&D strategy. Research institutes have new ideas and strong technical capabilities, while private companies understand the needs and limitations of the marketplace. If private companies are involved in the RD&D, it is much more likely that innovative technologies will be successfully commercialized and marketed. For example, collaboration between national laboratories and the private sector fostered the development, commercialization, and widespread adoption of electronic lighting ballasts, low-emissivity windows, and high-efficiency supermarket refrigeration systems in the United States (Geller and McGaraghan 1998). These efforts focused on both developing new technologies and improving linkages in the "innovation system."

A third lesson is that RD&D programs should support a wide range of technological concepts in the early stages of development and commercialization. This principal was followed in the successful wind turbine RD&D programs in Denmark and Germany, but was ignored in less successful RD&D programs in the United States and Sweden. The latter emphasized multimegawatt wind turbine designs chosen by government agencies rather than the private sector (Johnson and Jacobsson 2001, Loiter and Norberg-Bohm 1999). Uncertainty is high in the early stages of technological development. Thus it is important to "bet on" a variety of approaches and designs.

Greater international cooperation in energy technology RD&D has been recommended as a way to share costs and risks, speed up the learning process, and provide greater access to global markets (PCAST 1999). Partnering between industrialized and developing nations in energy RD&D could be particularly valuable for driving down the costs of new technologies such as fuel cells, bioenergy conversion technologies, or PV cells, as well as for deploying state-of-the-art clean energy technologies in the developing world. These countries offer enormous market opportunities, but there is often a need to subsidize early adopters and to tailor the design of clean energy technologies to the conditions in developing nations.

Financing

Financing at attractive interest rates can help to diffuse and build markets for energy efficiency and renewable energy technologies. Financing is facilitating the adoption of off-grid household solar PV systems in a number of countries, for example. In India, 10-year loans at low interest rates are offered by PV dealers as part of India's comprehensive renewable energy development program (see case study in Chapter 4). This has resulted in around 400,000 households using solar lanterns or PV lighting systems as of 2000 (Timilsina, Lefevre, and Uddin 2001).

Solar energy entrepreneurs who have access to a revolving loan fund financed solar PV systems for about 10,000 rural households in the Dominican Republic and Honduras as of 1999 (Verani, Nielsen, and Covell 1999). In the Dominican Republic, at least one company is employing a rental approach where it charges a monthly fee of up to US$20 for a small solar PV system but retains ownership and provides maintenance (Martinot et al. 2002). There is no down payment or other collateral required from the participant.

In Bangladesh, Grameen Shakti, an offshoot of the highly successful Grameen Bank, is marketing, supplying, and financing PV systems for rural households. This program focuses on women and provides loans for just two to three years (Hussain 2001). Many women use the light from solar systems to assist with income-generating activities. In addition to selling solar PV systems, Grameen Shakti trains purchasers in system maintenance.

Microcredit and revolving loan programs support solar entrepreneurs and the adoption of solar home energy systems in other countries including China, Indonesia, Mexico, Sri Lanka, and Vietnam. However, progress has been slow in some countries due to a range of problems, including costly or poorly performing solar systems, lack of a well-established marketing infrastructure, and high transaction costs (Martinot 2001, Miller and Hope 2000). This experience shows that microcredit schemes should be carefully designed to support good quality products and renewable energy businesses. Also, financing programs should minimize transaction costs (e.g., by making use of established financing entities such as rural credit cooperatives).

Financing is critical to the deployment of other renewable energy technologies besides solar PV systems. Low-interest loans, along with subsidies and development of local manufacturing, sales, and service capability, facilitated the adoption of small-scale wind power systems in Inner Mongolia. Around 140,000 household-size wind power systems, typically 100 to 300 watts in capacity, were adopted as a result of these policies (Martinot et al. 2002).

A number of countries have established financing programs for energy efficiency projects. Thailand, for example, created an energy conservation fund that receives $40 to $50 million per year through a small tax on the sales of petroleum products. But demand for this money was relatively limited during 1996 to 2001 due to poor marketing, overly bureaucratic procedures, and the broader economic crisis in Thailand. In response to these problems, a number of new energy efficiency financing and grants schemes were developed in 2001–02. The new schemes feature simplified procedures, improved marketing, use of commercial banks, and substantial financial incentives (Vongsoasup et al. 2002).

In North America, energy service companies (ESCOs) provide financing, technology, installation, and performance guarantees—a "one-stop" service for businesses or public agencies unable or unwilling to implement cost-effective efficiency measures on their own. A review of over 1,400 ESCO projects implemented over the past 20 years shows that nearly three-quarters of all projects have been in schools, hospitals, and government buildings—facilities that lack the capital and know-how to implement projects on their own. The typical cost of these energy efficiency projects is $500,000 to $2 million, and typical energy savings is 25 to 45 percent. The total ESCO market in the United States was about $2 billion as of 2000 and was increasing about 15 percent annually (Osborn et al. 2002).

A vibrant ESCO industry also exists in South Korea. As shown in Table 3-1, there were around 55 ESCOs operating in South Korea as of 1999. The total value of energy efficiency projects implemented by these ESCOs reached nearly $60 million per annum, and most of the funding was devoted to lighting, cogeneration, heat recovery, and industrial retrofit projects (Bang 2000). The development of the ESCO industry was facilitated through the creation of a special-purpose loan fund by the Korea Energy Management Corporation (KEMCO). The loan fund received its capital from the Korean government. KEMCO provides loans through commercial banks at attractive interest rates (5 percent per year) and loan terms (eight years). KEMCO also reviews and approves proposed energy efficiency projects (AID 1996).

Brazil is another country where many ESCOs are operating. But ESCOs are nonexistent or just starting to operate in other developing and transition countries (Sathaye and Ravindranath 1998). Consequently, the European Bank for Reconstruction and Development (EBRD) is providing both debt and equity financing to help develop a network of private ESCOs in some eastern European countries (Meyers 1998). Also, the World Bank and the Global Environmental Facility are funding projects to help establish or foster ESCOs in China and Hungary (Martinot and McDoom 2000).

TABLE 3-1

Development of the ESCO Industry in South Korea

Years	Number of ESCOs	Funding (million $)
1993	4	1
1994	6	6
1995	7	5
1996	7	7
1997	16	6
1998	29	28
1999	55	59

SOURCE: Bang 2000.

Financial Incentives

Financial incentives can help to get clean energy technologies established in the marketplace, encourage early adopters, and stimulate wider adoption. There are many examples of incentive programs for energy efficiency and renewable energy technologies, a few of which are mentioned here. Others are covered in more detail in the case studies in Chapter 4, including Brazil's ethanol fuel program, Denmark's wind power program, India's national renewable energy program, and energy efficiency efforts in California.

Many electric utilities in the United States provide incentives to households and businesses that purchase energy-efficient appliances, lighting products, motors, and the like. The cost per unit of energy savings is well below the cost of electricity from new power plants, meaning utilities (and their customers) find that it is less expensive to save energy than supply energy. The most successful programs feature substantial incentives as well as intensive marketing, and have achieved cumulative penetration rates of 50 percent or more (Nadel and Geller 1996). In addition, some U.S. utilities have used incentives to stimulate the development and commercialization of innovative technologies including "superefficient" refrigerators and clothes washers, new lighting products, and high-efficiency, environmentally friendly air conditioning systems (Lee and Conger 1996, Vine 2000).

Some utilities resist promoting energy efficiency by their customers since it will reduce their sales and revenues in the short run. Therefore, utility regulators have adopted energy efficiency program requirements and in some cases financial incentives to overcome this resistance. These financial incentives reward utilities based on the level of energy savings produced and/or cost effectiveness of their energy efficiency programs. For example, utilities in California, Massachusetts, New York, and Oregon were allowed to keep 8 to 27 percent of the net economic benefits produced by their energy efficiency programs during the mid-1990s (Stoft, Eto, and Kito 1995).

Tax credits are another type of incentive used to promote energy efficiency and renewable energy technologies in various countries. Tax credits were provided for energy efficiency improvements by households and businesses in the United States in the late 1970s and early 1980s. Studies of these tax credits found that most participants would have installed the measures in the absence of the incentives due to the small size of the tax credit and the inclusion of common efficiency measures (Geller 1999). Partly as a result of this experience, new tax credits have been proposed in the United States that are focused on innovative technologies such as hybrid and fuel cell vehicles, solar PV and biomass power systems, highly efficient appliances, and fuel cell cogeneration systems (Quinlan, Geller, and Nadel 2001).

Incentives for energy-efficient technologies can have limited impacts if not part of an overall market transformation strategy. For example, energy suppliers in the United Kingdom have provided incentives or free compact fluorescent lamps (CFLs) for many years.[1] It is estimated that nearly half the 17 million CFLs used in households as of 1998 were obtained through incentive or free distribution programs. But most of this dissemination was done outside of the retail market, and as a result there is still limited availability of CFLs in retail stores in the United Kingdom. Also, consumer awareness of the benefits of using CFLs is relatively low (Fawcett 2001).

There are many examples where financial incentives have been part of a market transformation effort. In the United States, incentives, promotion, and coordination with vendors greatly increased the availability and consumer acceptance of CFLs in the Pacific Northwest region in 2001 (NEEA 2002). In Mexico, incentives, bulk purchase, and promotion were used to introduce CFLs into the housing market in two cities (Friedmann 1998). Incentives and promotion were also used to build up the retail market for CFLs in Poland. In this case, incentives were paid to manufacturers to "buy down" the wholesale price and increase the impact of the limited funds that were available (Granda 1997). In both Mexico and Poland, incentives were offered for a limited period to increase awareness, prove commercial viability, increase product availability, and help lower costs (IEA 2001g, Martinot and Borg 1998). In the Netherlands, the government has subsidized the installation of small-scale gas-fired cogeneration systems, set attractive buyback rates for excess electricity sold to the power grid, and encouraged distribution utilities to invest in cogeneration systems. As a result, over 1,500 MW of small-scale gas-fired cogeneration systems were installed in the Netherlands during 1988–97. These systems cut national carbon dioxide emissions in 1997 by 3 percent (Strachan and Dowlatabadi 2002).

Renewable energy technologies other than hydropower are still not competitive with conventional fossil fuel technologies for providing power to the electricity grid, although the cost of electricity from renewable sources is rapidly declining. And renewable energy technologies, even if financed at attractive terms, are still too costly for many poor families not served by the electricity grid in rural areas of developing countries. But renewable energy technologies can be the least-cost option if all environmental and social costs are included in the comparison of resource options. Therefore, incentives are justified to help make renewable technologies affordable, establish and build markets, and move the technologies "down the learning curve." To achieve these goals, incentives should be well-designed and predictable. Incentives should be performance-based and scaled back as technologies mature and approach competitiveness.

Most industrialized countries and some developing nations provide financial incentives for renewable energy technologies through tax credits, low-interest loans, buyback rates above normal wholesale electric generation prices, or a combination of these options (Goldstein, Mortensen, and Trickett 1999). Feed-in laws—attractive fixed payments by utilities for electricity provided to the grid from renewable projects—have stimulated considerable renewable energy development in a number of European countries including Denmark, Germany, and Spain. This type of policy can make renewable sources such as wind power financially viable, while providing certainty for project developers.

In Germany, the Electricity Feed Law (EFL) paid renewable electricity generators 90 percent of the average retail price paid for electricity by end-users, leading to payments of around 10 cents per kWh (US$). Also, national banks offered low-interest loans for renewable energy investments, and some states provide additional subsidies (Moore and Ihle 1999). These incentives stimulated rapid growth in renewable-based electricity production, especially wind power. Wind power generation increased by over 700 percent between 1993 and 1999, and Germany led the world with over 6,000 MW of installed wind capacity by the end of 2000 (Pollard 2001).

Spain adopted an EFL in the early 1990s that made wind power and other renewable energy sources financially viable. Renewable energy producers are offered either a fixed price or an increment added to the average price of the market pool (Table 3-2). As a result, wind power capacity increased from about 50 MW in 1993 to 2700 MW in 2000, making Spain second or third in the world in terms of electricity generated by wind power (Aranda and Cruz 2000, Gipe 2000). Spain has set a goal of having renewable sources provide 12 percent of its total energy consumption by 2010 compared to about 6 percent in 2001 (IEA 2001e).

Electricity feed laws have come under attack in Europe by those in favor of "free markets" and open competition among all electricity sources. Early drafts of a renewable energy policy for the European Union (EU) placed quantity limits and a phase-out date on EFLs and weakened the overall goal of doubling renewable electricity supply in Europe by 2010 (Volpi 2000). These restrictions ignore the environmental problems caused by and subsidies provided to fossil fuels and nuclear power, as well as the importance of providing substantial and consistent incentives to advance renewable energy technologies.

The restrictions were strongly opposed by several countries, leading to a revised policy that allows continuation of national price support and incentive schemes. It also directs countries to set national targets consistent with the

TABLE 3-2

Prices Paid for Renewable Electricity Sources in Spain as of 2000[a]

Renewable Source	Bonus Price (Euro/kWh)	Fixed Price (Euro/kWh)
Wind power	0.0288	0.0626
Solar (< 5 kW)	0.3606	0.397
Solar (> 5 kW)	0.1803	0.216
Small hydro	0.0288	0.0626
Geothermal	0.0299	0.0636
Primary biomass[b]	0.0277	0.0615
Secondary biomass[b]	0.0256	0.0594

NOTES: [a]Renewable energy producers select one of two options—a varying price based on the average price of power in the market pool plus the bonus payment, or the fixed price.
[b]Primary biomass is dedicated energy crops; secondary biomass is agricultural and forest residues.
kW = kilowatts; kWh = kilowatt-hour .
SOURCE: Aranda and Cruz 2000.

overall EU renewables target and ensures that renewables are given fair access to the electricity grid (European Commission 2000). In early 2001, the EU Court of Justice upheld the right of individual countries to provide financial incentives and price supports for renewable energy sources.

In the United States, California began production-based incentive payments for new renewable energy projects in the late 1990s. The payments are set using a "supply auction" approach whereby project developers bid for incentives from a fixed incentive pool. As of 2000, 52 new renewable energy projects totaling over 500 MW in capacity were under development. The average incentive payment was 1.2 cents per kWh and payment duration was five years, significantly below predictions (Moore 2000). This represents the first significant renewable resource addition in California in over a decade. California also provides incentives to customers who choose to purchase renewable-based "green power" from retail providers. These incentives started at 1.5 cents per kWh in 1999 but were lowered when demand exceeded expectations (Moore 2000).

In Japan, the federal government provides substantial capital subsidies for rooftop PV systems. The budget for this program was steadily increased during the 1990s and reached nearly 18 billion yen (US$150 million) as of 2000 (Table 3-3). Also, utilities are required to pay 15 to 19 cents per kWh for PV power fed into the utility grid as well as about 10 cents per kWh for wind power (Shoda 1999). As a result of these policies, Japan led the world with over 50,000 grid-connected solar-powered homes aggregating to over 350 MW of PV capacity as of 2001 (Rever 2001, Sawin 2002).

Building the market for residential PV systems in this manner has greatly

TABLE 3-3

Residential Solar PV Incentives Program in Japan

	1994	1995	1996	1997	1998	1999[a]	2000[a]
Number of homes	540	1,100	2,000	5,700	6,400	17,400	25,700
Installed capacity (MW)	2	4	8	20	24	64	96
Incentive limit (yen/W)	900	850	500	340	340	329	150–270[b]
Incentive budget (billion yen)	2	3.3	4.1	11.1	14.7	16.0	17.8

NOTES: [a]Values are based on number of applications in 1999 and 2000.
[b]The incentive limit was 270 yen/W at the beginning of 2000 but was reduced to 180 yen/W and then 150 yen/W during the year.
SOURCE: New Energy Plaza 2001.

reduced the cost of PV systems in Japan. The total installed cost dropped from about $30 per watt when the program began in 1993 to about $8 per watt in 1998 (IEA 2000e). This enabled the Japanese government to reduce its subsidy from 900 yen per watt in 1994 to just 150 yen per watt by the end of 2000 (New Energy Plaza 2001). Thus Japan was successful in using "market pull" to drive down the cost of an emerging renewable energy technology.

In 1997, Japan adopted ambitious goals of installing 400 MW of PV capacity by 2000 and 4,600 MW by 2010, including installations outside of the residential incentives program (Oliver and Jackson 1999). Kyocera, Sharp, Sanyo, and other Japanese PV manufacturers have increased output to meet the growing demand. These companies now lead the world in PV production (Maycock 2001).

Germany launched a program in 1999 to install 100,000 grid-connected PV systems within six years. The program includes 10-year, low-interest loans and attractive buyback rates to stimulate installation of PV systems (Moore and Ihle 1999). Also, Germany revised its payments under the EFL to achieve a target of 10 percent renewables in the electricity supply mix by 2010. The new policy includes differentiated payments for various renewable energy sources and also allows utilities as well as independent producers to receive the payments. Owners of solar PV systems receive DM 0.99/kWh (US$0.45/kWh) for electricity supplied to the grid. With these incentives, adoption of building-integrated PV systems rapidly expanded with over 30,000 systems installed or under development as of 2001. Total PV capacity in Germany increased from less than 20 MW in 1995 to nearly 115 MW as of 2000 (Weiss and Sprau 2002).

Programs such as the Japanese and German rooftop PV programs have led to very rapid growth in grid-connected PV systems. These systems accounted for about 40 percent of PV installations in 2000 compared to just 3 percent in 1993 (Maycock 2001). The grid-connected PV market in Japan alone was 15 times greater than that in the United States as of 2000 (Maycock 2001).

The Global Environmental Facility (GEF) was created to help developing countries pay the incremental costs for technologies that offer significant global environmental benefits. The GEF approved grants of $706 million for 72 energy efficiency and renewable energy projects in 45 countries from 1991 to 1999, although not all these projects got off the ground (Martinot and McDoom 2000). While GEF and World Bank support for renewable energy is on the rise, only about 8,000 solar home systems and less than 100 MW of grid-connected renewables were financed and installed as of 1999 (Martinot 2001). Establishing vibrant markets for renewable energy technologies has proven to be very challenging in many (but not all) developing countries.

Renewable energy markets in some developing countries have been hurt by high subsidies and donor-driven programs. In some cases, developing countries were required to purchase subsidized equipment from the donor country, inhibiting development of local manufacturing capability. This restriction delayed manufacturing of large-scale wind turbines in China, for example (Lew and Logan 2001).

High subsidies and donor-driven programs can also lead end users to expect free or heavily subsidized systems, which undermines the development of commercial markets. In the case of Zimbabwe, many local enterprises entered the solar PV market in response to a heavily subsidized donor-based program. Once the program ended, these enterprises collapsed, and households were left without any service infrastructure (Martinot et al. 2002). The program hindered rather than helped to build an orderly, sustainable market for solar PV systems in Zimbabwe.

On the other hand, India has demonstrated that carefully designed incentives along with attractive financing, renewable energy business development and support, and a long-term government commitment can succeed in establishing healthy markets for renewable energy technologies in developing nations (see case study in Chapter 4).

Pricing

Many but not all nations heavily tax gasoline and diesel fuel because of their significant social and environmental costs, especially countries heavily dependent on oil imports (Fig. 3-3). Some European nations have increased their traditionally high gasoline taxes in recent years. Germany, for example, increased taxes on gasoline and diesel fuel substantially during the period from 1990 to 1999 (IEA 2000c). The United Kingdom increased fuel taxes 5 percent per year above

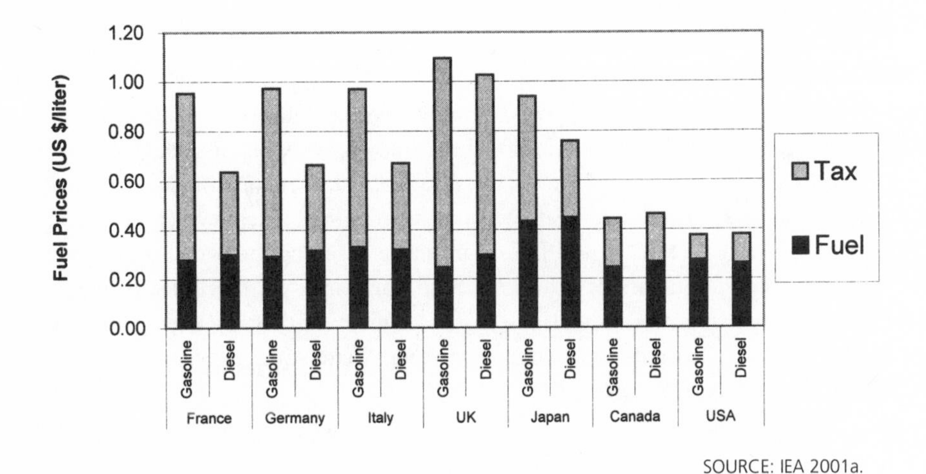

SOURCE: IEA 2001a.

FIGURE 3-3: Gasoline and diesel fuel prices and taxes in OECD nations, March 2001.

inflation for a number of years, with some of the additional revenue used to offset a reduction in taxes on smaller and more fuel-efficient vehicles. Due in part to high fuel taxes, passenger vehicle fuel use per capita in Western Europe is just one-third to one-half as great as in North America (IEA 1997d).

Further energy tax reform has been introduced or is under discussion in a number of Organization for Economic Cooperation and Development (OECD) countries in the context of efforts to cut carbon dioxide (CO_2) emissions and meet the targets in the Kyoto Protocol. Five countries—Denmark, Finland, the Netherlands, Norway, and Sweden—enacted modest taxes on carbon emissions or fossil fuels in the early 1990s as part of a revenue-neutral tax shift strategy (Roodman 1998). Taxes on carbon emissions or fuel use are offset by reductions in other unpopular taxes such as personal income, retirement, or employment taxes. In some cases, a portion of the tax revenue is targeted for funding energy efficiency and renewable energy RD&D or incentive programs. The carbon or fossil fuel taxes adopted in Europe are estimated to have reduced CO_2 emissions by about 1.5 percent in the Netherlands, 3 to 4 percent in Norway, 4 to 5 percent in Finland, and about 5 percent in Denmark (Vehmas et al. 1999).

In Sweden and Denmark, energy and carbon taxes, along with tax breaks on biofuels, contributed to large increases in bioenergy use for combined heat and power as well as district heating. Since the introduction of carbon taxes and biofuel tax breaks, district heating using biofuels increased by 70 percent in Sweden while district heating using coal declined by 50 percent (IEA 2000e). Biomass is now the predominant fuel for district heating in Sweden. Furthermore, biomass provided about 91 TWh or 19 percent of total electricity supply as of 1997 (Jacobsson and Johnson 2000).

The United Kingdom adopted an innovative tax shift policy in 2000. Starting in 2001, businesses paid an additional tax, known as a Climate Change Levy, on fossil fuels and electricity. The tax does not apply to new renewable sources and energy-efficient combined heat and power generation. The new tax was accompanied by a reduction in health insurance taxes, making the overall scheme revenue neutral to businesses as a whole. Also, companies in the United Kingdom's most energy-intensive industrial sectors that meet challenging energy efficiency or CO_2 emissions reduction targets receive a rebate equal to 80 percent of the new energy tax (IEA 2001b).

Carbon taxes, if large enough, can increase energy efficiency and promote greater use of cleaner energy sources, thereby reducing CO_2 emissions. An analysis for Japan, for example, suggests that an $80 per ton carbon tax could lower CO_2 emissions in 2040 by 9 percent, whereas a $160 per ton tax would lower emissions by 20 percent (Nakata and Lamont 2001). A carbon tax is more effective for reducing CO_2 emissions than an equivalent energy tax because it tends to shift power generation from coal to natural gas and renewables.[2] Also, dedicating a portion of energy or carbon tax revenue to energy efficiency efforts can significantly increase energy savings and emissions reductions compared to those resulting from the tax by itself (Geller, DeCicco, and Nadel 1993).

Differential fuel taxes have been adopted in a number of developing countries to reduce oil import dependence and improve urban air quality. In Brazil, ethanol fuel and vehicles that run on it are taxed less than gasoline and gasoline-fueled vehicles (see case study in Chapter 4). Other countries have eliminated or reduced taxes on compressed natural gas (CNG) to encourage natural gas vehicles. Argentina, for example, eliminated taxes on CNG and constructed a network of CNG filling stations starting in the mid-1980s. As a result, Argentina has the largest CNG car fleet in the world. Around 425,000 CNG vehicles operate in Argentina, about 10 percent of all cars on the road (Suarez 1999).

Energy price subsidies have been eliminated or reduced in many developing and transition countries in recent years (Reid and Goldemberg 1998). Russia, for example, reduced fossil fuel subsidies by about two-thirds between 1991 and 1996. China cut its fossil fuel subsidies by over 50 percent. And Mexico increased the price of gasoline by 52 percent and fuel oil by nearly 100 percent between 1988 and 1996. Overall, total fossil fuel subsidies declined 45 percent in 14 major developing countries from 1990–91 to 1995–96 (Sathaye and Ravindranath 1998).

Reducing or eliminating energy price subsidies can improve energy and economic efficiency, but it can also have an adverse impact on low-income households. Energy costs account for as much as 40 percent of a typical Ukrainian household's expenditures, for example, and cuts in energy subsidies have significantly increased this expense (IEA 2001g). Among lower-income households in rural areas of India, energy costs represent as much as 50 percent of nonfood expenditures (Cecelski 1995). As price subsidies are phased out in poorer nations, it is reasonable to assist low-income households through reduced energy rates as well as free or low-cost energy efficiency and renewable energy services.

Requiring utilities to purchase power from renewable energy suppliers and/or cogenerators at their avoided costs is one type of pricing policy. In the United States, the Public Utility Regulatory Policy Act of 1978 (PURPA) included this policy as well as requirements that utilities provide standby and backup power to cogenerators at reasonable rates. As a result, cogeneration capacity in the United States increased from about 10,000 MW in 1980 to 46,000 MW in 1996, with cogenerators accounting for about 9 percent of total power generation in 1996 (Bluestein and Lihn 1999). But with the advent of competitive wholesale markets in the mid-1990s, PURPA became much less effective, and the amount of cogeneration leveled off.[3]

Net metering is another pricing policy that can be used to encourage renewable electricity generation. Net metering means that customers who produce more electricity than they can use feed the extra electricity into the grid and run their meter backward, thereby selling power to the utility at the retail price rather than a wholesale price. In the United States, over 30 states adopted net metering as of 2001, although most include eligibility limits based on the size of the renewable facility (Clemmer et al. 2001).

Some countries heavily tax energy-intensive goods or activities to encourage more fuel-efficient behavior. For example, the tax on new cars purchased in Denmark is equal to 180 percent of the vehicle price (IEA 2000c). In addition, Denmark recently modified its vehicle registration tax to make it proportional to both fuel economy and weight, with a considerable tax break for cars with a fuel intensity under 4 liters per 100 kilometers. These policies, in combination with high fuel taxes, have resulted in very low car ownership and a relatively efficient vehicle fleet, especially considering the nation's high income. Transport energy use in Denmark rose only about 5 percent from 1990 to 1997, compared to nearly a 12 percent increase in the EU as a whole (IEA 2000c).

Singapore is another country that has heavily taxed personal vehicle ownership and usage. These policies, along with substantial investment in public transport infrastructure and coordination of land-use and transport development, have resulted in very low transport energy use considering Singapore's level of affluence (Sperling and Salon 2002).

Municipalities can use their taxing authority to promote more energy-efficient urban development or transport. San Francisco, for example, has implemented a "transit impact development fee" on new office construction or renovation in the downtown area. The funds collected help support the operation of public transit systems in the city. Los Angeles charges city workers a parking fee if they drive to work solo and uses the funds to support car pooling, van pooling, and public transit use. Other cities in California charge developers impact fees if they build outside of the urban center to discourage urban sprawl (ICLEI 2000b).

Voluntary Agreements

Voluntary agreements between governments and the private sector can result in energy efficiency improvements in either manufactured products or energy use. In Thailand, a voluntary agreement among the government, a utility, and lighting manufacturers succeeded in phasing out production of less efficient fluorescent lamps by 1995. The government supported the transition from 40 watt T12 lamps to more efficient 36 watt T8 lamps through an extensive public awareness and educational campaign. It is estimated that this initiative reduced peak demand in Thailand by about 630 MW as of 2000 (Birner 2000).

In Europe, auto manufacturers and the European Commission completed a voluntary agreement in 1998 whereby manufacturers pledged to achieve a new car fuel economy target of 140 grams of CO_2 emission per kilometer (g/km) by 2008. This value is 25 percent below the average of 186 g/km for new cars sold in Europe in 1995 (ACEA 1998). This nonbinding agreement was modeled on a voluntary agreement by German car manufacturers to reduce the fuel use and CO_2 emissions of new cars by 25 percent between 1990 and 2005 (IEA 2000c). The new car fleet sold in Europe in 2001 averaged 164 g/km, indicating that manufacturers were on track for meeting the 2008 goal (ACEA 2002).

Voluntary agreements between governments and manufacturers have also been used to increase the energy efficiency of water heaters, clothes washers, dishwashers, and electronic appliances sold in Europe. The agreements negotiated and signed by the European Commission contributed to about a 20 percent reduction in the energy consumption of new clothes washers and dishwashers, and a 25 to 35 percent reduction in the standby power consumption of TVs and VCRs (Bertoldi, Waide, and Lebot 2001).

The Netherlands successfully implemented voluntary agreements with industries for reducing manufacturing energy intensity (see case study in Chapter 4).

Germany also adopted voluntary agreements with major industries and utilities to achieve an overall 20 percent reduction in energy use and CO_2 emissions between 1990 and 2005 (Gummer and Moreland 2000). Nineteen industry and utility associations, representing about 70 percent of industrial energy consumption and 99 percent of public power generation, signed such agreements. In return, industries hope that the German government will avoid adopting more onerous regulatory or tax measures. Monitoring indicates that industries were on track for meeting the targets as of 1997, although data collection and verification need to be improved (Gummer and Moreland 2000).

Voluntary agreements have the advantage that they can be easier and faster to adopt than mandatory regulations. On the other hand, voluntary agreements usually are not legally binding and consequently the targets may not be achieved. Furthermore, there is no guarantee that companies will make strong commitments that compel them to go beyond "business-as-usual." Voluntary agreements tend to be more effective when companies fear that they will face tougher policies, such as taxes or regulations, should they fail to establish and meet strong voluntary targets (Newman 1998, Price and Worrell 2000). Likewise, companies will tend to make and follow through on their voluntary commitments if they think it can provide a strategic advantage. Successful voluntary programs also require thorough monitoring and evaluation as well as recognition of companies that achieve significant results.

Voluntary commitments can also be made by governmental authorities. In the United States, the state of New Jersey has voluntarily committed to reduce its greenhouse gas emissions by 20 million tons of CO_2 equivalent by 2005 (NJDEP 2000). This is 3.5 percent below emissions in 1990 and a 13 percent reduction compared to projected emissions in 2005 in the absence of the policy. New Jersey plans to meet this goal through a combination of energy efficiency, renewable energy, transportation, recycling, and landfill gas recovery initiatives as well as by encouraging voluntary reductions by businesses and industries in the state. In particular, New Jersey adopted a 2 percent surcharge on electricity sold in the state in order to fund energy efficiency and renewable energy programs (Kushler and Witte 2001).

A number of cities have voluntarily committed to increase their energy efficiency, expand renewable energy use, and/or reduce their CO_2 emissions. Toronto, Canada, has pledged to reduce its CO_2 emissions 20 percent by 2005. To achieve this goal, Toronto established a revolving loan fund known as the Toronto Atmospheric Fund. The Fund is being used to retrofit city-owned buildings and municipal lighting, expand the district heating and cooling system in the downtown area, capture methane from a municipal waste site, support renewable energy projects, and improve mass transit and nonmotorized transportation infrastructure (Jessup 2001). Many other cities have made similar commitments and developed emissions reduction strategies through the Cities for Climate Protection Campaign organized by the International Council for Local Environmental Initiatives (ICLEI 2000a).

In 2001, Seattle, Washington, adopted a goal of zero net greenhouse gas emissions by its municipally owned electric utility. The utility sold its share of a coal-fired power plant and plans to reduce or offset emissions from its remaining fossil fuel resources—600,000 metric tons of CO_2 each year. The utility is expanding energy efficiency programs and renewable energy acquisition and has issued its first request for carbon offset projects (Seattle 2001).

Regulations

Regulations can be adopted to increase energy efficiency or overcome barriers to the adoption of renewable energy devices or cogeneration systems. Minimum efficiency standards either remove the least efficient products from the marketplace or require that all new products meet a certain average efficiency level. By making more efficient products the norm, economies of scale occur and the cost premium for energy savings is reduced. In addition, efficiency standards can provide assured markets for innovative technologies.

Many countries have adopted minimum efficiency standards for mass-produced goods such as cars, domestic appliances, heating and cooling equipment, motors, and lighting products. In the United States, national efficiency standards were first adopted on major residential appliances in 1987. National efficiency standards were extended to heating, air conditioning, and lighting products used in commercial buildings as well as to motors in 1988 and 1992. These standards have saved a significant amount of energy, money, and pollutant emissions (see case study in Chapter 4). In Europe, mandatory efficiency standards apply to refrigerators and freezers only.

Japan enacted mandatory standards at various times to improve the energy efficiency of vehicles as well as appliances (IEA 2000b). Standards resulted in a 20 percent improvement in the efficiency of new refrigerators and a 17 percent improvement in the efficiency of air conditioners in the early 1980s. In 1993, new targets were established for cars, air conditioners, fluorescent lamps, televisions, and some types of office equipment. The program was updated again in 1998, with the new efficiency targets based on the most energy-efficient products available at the time (IEA 2001c). This is known in Japan as the "Top-Runner" approach (Table 3-4). The Top-Runner program is expected to reduce overall CO_2 emissions in Japan 6 percent by 2010.

TABLE 3-4

Energy Efficiency Target Levels of the Japanese "Top-Runner" Program

Product	Standard Level	Efficiency Improvement (%)[a]	Target Date (year)
Vehicles			
Gasoline cars	6.4–21.2 km/L	23	2010
Gasoline light trucks	9.3–20.2 km/L	13	2010
Diesel cars	8.7–18.9 km/L	15	2005
Diesel light trucks	9.9–17.7 km/L	6.5	2005
Refrigerators	varies by volume	30	2004
Air conditioners	2.5–3.6 COP	15	2007
Heat pumps	2.8–5.3 COP	63	2007
Fluorescent lamps	49–86 lumens/W	17	2005
Televisions	varies by size	16	2003
VCRs (standby)	1.7–4.0 W	59	2003
Photocopiers	varies by copy rate	30	2006
Computers	varies by size/speed	83	2005
Magnetic disk drives	varies by speed	78	2005

NOTES: [a]Improvement from average efficiency as of 1997. km/L = kilometers per liter; W = watt; COP = coefficient of performance.
SOURCES: IEA 2000b, 2001c.

Developing countries that have adopted efficiency standards include South Korea, the Philippines, and Mexico. South Korea launched a standards and labeling program for appliances and lamps in 1992; standards have been strengthened over time and equipment efficiency has significantly improved (Egan and du Pont 1998). The Philippines adopted minimum efficiency standards for room air conditioners and refrigerators and is considering standards for motors (Wiel et al. 1998). Mexico enacted mandatory efficiency standards on refrigerators, air conditioners, and motors, initially at modest levels of efficiency but later equivalent to U.S. standards (Friedmann 1998).

Most industrialized countries and some developing nations have adopted building energy codes that specify minimum energy efficiency requirements for new residential and commercial buildings. Building energy codes can be either prescriptive or performance based, or a combination of the two. Prescriptive standards mandate certain technologies or system configurations. Performance standards require a certain level of energy performance such as a minimum insulation value in walls or a maximum amount of installed lighting power per unit of floor area. Performance standards provide building designers and owners greater flexibility but can be more difficult to monitor and enforce.

Building codes can be an important energy savings strategy over the long run, especially if building codes are updated periodically as building technologies and design strategies evolve. The United Kingdom, for example, raised its national building energy efficiency requirements by 25 to 35 percent, effective in 1995. The U.K. raised these standards again in 2001, with the changes taking effect in 2002 (IEA 2001b). Building energy codes tend to be most effective if accompanied by training of building designers and construction firms, promotion of innovative and low-cost compliance techniques, and rigorous code enforcement (Halverson et al. 2002).

In developing countries where the building stock is expanding rapidly, building energy codes can have a significant impact on energy use within 10 to 20 years. But it can also be difficult to enact and implement mandatory building energy codes in developing countries due to cost concerns, opposition from the construction industry, and weak enforcement (Flanigan and Rumsey 1996). In South Africa, for example, it has been very difficult to build thermally efficient new homes as part of the national housing construction program because of opposition to mandatory standards and the fact that these houses would have a slightly higher up-front cost (Spalding-Fecher, Williams, and van Horen 2000). The Philippines enacted a building energy code for new commercial buildings in 1994, but the code has not been enforced or followed (Wiel et al. 1998). Building code enforcement has also been lacking in China.

Regulations can overcome barriers to the adoption of cogeneration and distributed generation technologies, specifically with regard to providing uniform, reasonable technical specifications for connecting to the power grid. In the United States, a number of states have enacted such standards, and federal interconnection standards were under development as of 2001 (Meyers and Hu 2001). Efficiency standards can also be enacted to stimulate construction of state-of-the-art power plants. In the United States, the state of Oregon has set a CO_2 emissions standard on new power plants. The standards require new plants to be combined-cycle gas turbines with an electrical efficiency of at least 57 percent. However, a utility or developer can purchase CO_2 emissions offsets or pay into a fund that invests in carbon reduction projects if they choose to build a less efficient plant (Sadler 1999).

Performance standards and product certification can improve the quality of solar PV systems. In Kenya where on the order of one-third of solar PV home systems perform poorly, mandatory performance standards and product labeling have been proposed (Duke, Jacobson, and Kammen 2002). Experience in Indonesia, Sri Lanka, and China demonstrates that it is important to strike a balance between equipment performance, promoting local production, and product affordability (Martinot, Cabraal, and Mathur 2000). Standards and certification programs should contribute to consumer satisfaction and market sustainability, but not excessively stifle the growth of solar PV markets.

Regulations can also be used to influence behavior and limit use of energy-intensive devices such as cars. However this requires caution to avoid potential unintended consequences. Mexico City, for example, tried banning use of cars one day each week to reduce urban congestion and air pollution. But public transit systems in Mexico City were not improved in conjunction with the car restrictions, so many residents bought an extra car, often an older and highly polluting one. The car restriction ended up worsening air pollution, not improving it (Sheehan 2001).

Information Dissemination and Training

Many energy authorities educate consumers and businesses regarding energy efficiency and renewable energy measures through energy audits, labels, advertisements, and distribution of printed materials. The U.S. Green Lights and Energy Star® commercial buildings programs provided training, information, analytical tools, and recognition once companies agreed to audit their facilities and install more efficient lighting and other efficiency measures wherever cost-effective. As of 2000, participants cut their electricity use by nearly 32 billion kWh per year, thereby preventing about 6 million metric tons of carbon emissions annually (EPA 2001). The success of this effort is directly attributed to the commitment required of companies before they are eligible to receive assistance and recognition.[4]

Appliance testing and labeling programs can be useful for informing consumers about the relative energy efficiency of different products. This strategy enables labeling and promotion of the top-rated products available at any particular time, as has been done in the United States, Brazil, and elsewhere. Testing and labeling can be very cost-effective for governments since all of the investment in more efficient products is made by the private sector and paid for by consumers. By identifying high-efficiency products, standardized testing also provides the foundation for incentive programs and minimum efficiency standards.

In the EU, manufacturers are required to test and label the energy performance of refrigerators, freezers, clothes washers, and other household appliances. The label rates each product from "A" to "G" based on its energy efficiency, making it easy for consumers to identify efficient products. The combination of labeling and minimum efficiency standards that took effect in 1999 reduced the average electricity consumption of new refrigerators and freezers sold in the EU by 27 percent between the early 1990s and 1999 (Waide 2001). Furthermore, it is estimated that for each Euro spent on labeling, consumers will save about

100,000 Euros on their electricity bills (Wiel and McMahon 2001). However, because of the significant efficiency improvements already made, both the labeling scheme and the efficiency standards need to be updated (Waide 2001).

In the United States, the Energy Star® product rating and labeling program is informing consumers of high-efficiency appliances, office equipment, lighting products, and other devices. The program also works with manufacturers to increase the availability of efficient products. As of 2000, more than 600 million products with the Energy Star® label were purchased. These products are saving about 42 billion kWh per year, worth about $3 billion in reduced energy bills (Table 3-5). The electricity savings also translate to about 9 million metric tons of avoided carbon emissions (EPA 2001). The program has had the biggest impact on the energy efficiency of personal computers, monitors, and other types of office equipment.

Training can help to ensure that energy-efficient products are installed and used properly. In the United States, for example, training facility managers in commercial buildings is critical for realizing the savings potential from energy management and control systems (Dodds, Baxter, and Nadel 2000). Also, training contractors who install air conditioning systems can increase the number of units that are sized and installed properly. If this is not done, high efficiency air conditioners will not save as much energy as they could (Neme, Proctor, and Nadel 1999).

Training is an important element of successful energy efficiency and renewable energy programs in China and India (see Chapter 4). Training has also been critical to energy efficiency and renewable energy efforts in Africa. For example, the improved cookstove program in Kenya involved training stove artisans to

TABLE 3-5

Savings in 2000 from Energy Star® Labeled Products in the United States

Product	Start Year	Energy Savings (Billion kWh)	Avoided Carbon Emissions (MMT)
Computers and monitors	1993	24.5	5.0
Printers	1993	6.0	1.2
Copiers		1.1	0.2
Other office equipment	1995–97	3.4	0.7
TVs, VCRs, audio equipment	1998–99	2.0	0.4
Lighting products	1995–97	3.9	0.8
Other products	1996	1.5	0.3
TOTAL		42.4	8.8

NOTES: kWh = kilowatt-hour; MMT = million metric tons.
SOURCE: EPA 2001.

make improved, more fuel-efficient stoves. Over 200 artisans and small-scale businesses have sold around 700,000 improved stoves in Kenya where they are used in the majority of urban homes (Kammen 1999, Karekezi and Ranja 1997).

Experience shows that information dissemination or training alone may result in limited energy savings. During the 1980s, for example, utilities in the United States were required to provide free energy audits to their residential consumers. After six years, about 7 percent of households participated in the program and energy savings averaged just 3 to 5 percent per participant (Nadel and Geller 1996). Likewise, energy audits and other educational efforts during the 1980s resulted in limited energy savings in Brazil (Geller 1991). Information and training tends to be more effective when targeted to decision makers and provided at the time when a building is being constructed or an energy-intensive product is being purchased.

Information dissemination also tends to be more effective when it is combined with other policies such as financing, incentives, voluntary agreements, or regulations. In China, education and training provided by energy efficiency service centers at the provincial and local levels has been successful in conjunction with RD&D, financing, incentives, and regulations (Sinton, Levine, and Qingyi 1998). In Thailand, a public awareness campaign for energy-efficient fluorescent lamps was successful, carried out in conjunction with a voluntary agreement with lamp manufacturers (Birner 2000). And in the United States, education and training of builders and building code officials has been effective, carried out in conjunction with adoption of more stringent building energy codes (Geller and Thorne 1999).

Information dissemination and training often are key components of efforts to disseminate solar water heating, PV power, and other renewable energy devices in developing countries (Kammen 1999). China in particular has established effective networks of local information dissemination and training centers as part of its strategy to promote biogas digesters, small-scale wind turbines, and other renewable energy technologies (Martinot et al. 2002). Likewise, India has established regional training, promotion, and service centers as part of its biogas, solar PV, and other renewable energy programs (see case study in Chapter 4). But once again, information dissemination and training are part of comprehensive renewable energy programs in both China and India.

Procurement

Bulk purchases by government authorities or the private sector can help establish initial markets for clean energy technologies. Governments—federal, state, and municipal—buy large numbers of lamps, air conditioning equipment, vehicles, appliances, and other mass-produced goods. Routinely purchasing energy-efficient products will save the buyers money on a life-cycle basis as well as help to establish and build the market for innovative technologies.

Government procurement has been used in this manner in the United States. Starting in 1993, U.S. federal agencies were required to buy Energy Star® personal computers and other types of Energy Star® office equipment. This was one of the key factors that led to widespread production of Energy Star® personal computers by manufacturers worldwide (Thigpen et al. 1998). An executive order issued in 1999 requires federal agencies to purchase Energy Star® products when purchasing lighting products, air conditioners, appliances, and other energy-consuming products (Wiel and McMahon 2001).

The New York Power Authority (NYPA) is using bulk procurement to replace older refrigerators in 180,000 apartments occupied by low-income households in New York City. NYPA is reimbursed by the city's housing authority, which pays the electric bills for these households. In response to the bulk purchase offer, one appliance manufacturer developed a new apartment-sized refrigerator model that consumes about 60 percent less electricity than typical models found in these apartments. Older refrigerators are collected and their materials are recycled as part of this program (Kinney and Cavallo 2000). In addition, utilities and housing agencies in other cities have undertaken similar bulk purchases.

In Sweden, the National Board for Industrial and Technical Development (NUTEK), now known as the Swedish National Energy Administration, organized the bulk procurement of a number of high-efficiency appliance, lighting, and building technologies. Purchase commitments were made by both public agencies and private firms, with manufacturers competing for the procurement. In a number of cases, the bulk procurement was complemented with information and promotion efforts, labeling, voluntary standards, and rebates.

These efforts stimulated the introduction of high-efficiency refrigerators, clothes washers and dryers, high-performance windows, more efficient computer equipment, and electronic lighting ballasts in Sweden (Neij 2001). For example, a procurement and solicitation for high-efficiency refrigerator-freezers led to the introduction of a model that used 50 percent less electricity than the average for other models in the market at the time. The cost to the Swedish government for these procurement efforts was relatively low since it mainly played a facilitating role.

One of the first Swedish procurement programs was for high-frequency electronic ballasts. A total of 46,000 ballasts were purchased, accelerating the introduction and successful market development for this energy-saving technology (Fig. 3-4). By the mid-1990s, the cost of high-frequency ballasts dropped

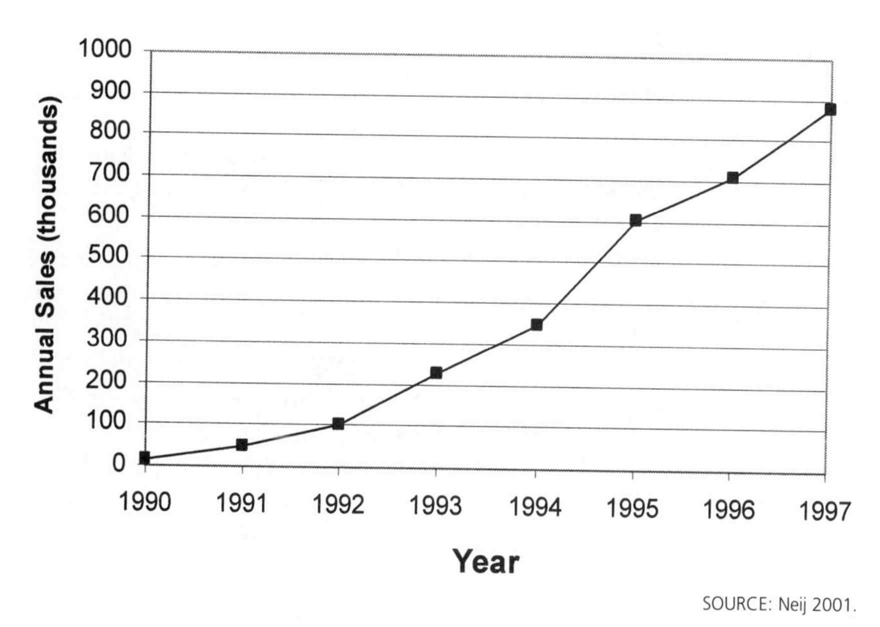

SOURCE: Neij 2001.

FIGURE 3-4: Development of the market for high-frequency electronic ballasts in Sweden.

significantly, and market share increased to 60 to 70 percent of new ballast installations. The economic benefits of this program were estimated to equal about 20 times the cost (Neij 2001).

Some developing countries have used bulk procurement to help establish markets for energy efficiency measures. The national utility in Mexico purchased CFLs in bulk and sold them to consumers in two cities. By purchasing in bulk, the utility obtained high-quality CFLs at a substantial discount from manufacturers. The utility financed the purchase of the lamps by consumers and also provided a subsidy. Over 2.5 million CFLs were sold from 1995 to 1998 (Friedmann 2000). Initially it was feared that this program would take sales away from commercial vendors of CFLs. But by increasing the awareness of CFLs, the overall market was increased, and distributors and retailers benefited.

In Thailand, a utility bought CFLs in bulk and sold them through convenience stores without a subsidy. Bulk purchase lowered the price of CFLs by about 40 percent and increased their availability in retail outlets (Martinot and Borg 1998). About 900,000 CFLs were sold in this manner as of 2000.

Government procurement has focused mainly on energy-efficient technologies so far, probably because these products tend to be cost effective. However, governments own large numbers of buildings that can utilize building-integrated renewable technologies or fuel cell cogeneration systems. Governments can purchase "green power" with a high renewable energy content for use in their own facilities, as well as coordinate the purchase of renewable energy technologies or green power by large numbers of households or businesses. Also, government vehicle fleet purchases have been suggested as a strategy for launching hydrogen fuel cell vehicles in the marketplace (Ogden, Williams, and Larson 2001).

Government procurement of renewable energy technologies or fuel cells will require subsidies, at least initially. But government procurement can expand the market for and reduce the cost of renewable energy technologies such as solar PV systems over the long run (Box 3-1). Governments are beginning to recognize this opportunity and purchase renewable energy devices and green power on a significant scale. The state of New York, for example, has committed to purchase green power from renewable sources or fuel cells for 10 percent of the electricity used in state facilities by 2005 and 20 percent by 2010 (NYS 2001). The city of

BOX 3-1

The Power of Government Procurement: Potential Solar Photovoltaic Cost Reductions

Government procurement could expand the market for and reduce the cost of photovoltaic (PV) cells in the United States. If federal facilities obtained just 1 percent of their electricity from PV cells, they would utilize about 335 MW of PV capacity. This is three times the total PV capacity installed in the United States from 1994 to 2000 (Maycock 2001). Assuming an average PV system cost of $5 to $7 per watt, adding 335 MW of PV capacity would represent about $1.7 to $2.3 billion of investment in PV systems. The investment could be spread out over a number of years.

Total U.S. PV production (including exports) over the past decade equaled about 360 MW. Historically, the cost of PV modules has dropped about 20% for each doubling in cumulative production (EIA 1998, Maycock 2000). Thus, adding 335 MW of capacity via government procurement could significantly reduce the cost of producing PV systems. If state and local governments purchased an equivalent amount of capacity, the combined effect could reduce PV system costs by as much as $1.50 to $2.00 per watt. A cost reduction of this magnitude would enable PV systems to compete in many more applications in the United States and elsewhere.

◉ ◉ ◉

Chicago and nearby municipalities have committed to buying 20 percent of their electricity from renewable sources by 2006. This commitment should result in 80 MW of renewable energy capacity in this metropolitan area.

Market Reforms

Market reforms that introduce greater competition and reward higher levels of energy efficiency can help to overcome some of the barriers to sustainable energy development. A number of countries, both industrialized and developing, have restructured their power sectors to stimulate greater competition, privatization, and investment. In some cases this has led to increased installation of technologies such as combined-cycle natural gas–fired power plants, which in turn has increased the efficiency of electricity generation and cut pollutant emissions. The United Kingdom is one country where this has occurred (see case study in Chapter 4).

Argentina provides another example of market reforms enhancing energy and economic efficiency. The average efficiency of thermal power plants in Argentina increased from about 25 percent in 1989 to about 40 percent by 1998 following utility privatization and adoption of policies to increase competition among generators (Dutt, Nicchi, and Brugnoni 1997). These reforms also led to greater power plant availability, reduced distribution losses, and lower electricity prices. However, no specific steps were taken to increase the efficiency of electricity use in conjunction with these market reforms.

As discussed in Chapter 2, deregulation and privatization can reduce utility support for end-use energy efficiency programs, renewable energy development, and expansion of utility service to low-income and rural households. As a consequence, it is important to adopt specific policies to protect these "public goods" as part of utility restructuring. A renewable energy market obligation is one such policy, and is presented in the next section.

Allowing product differentiation and marketing of green power is another policy that can help to advance renewable energy in a more competitive market environment.[5] In the state of Colorado, for example, over 17,000 households and businesses have agreed to pay a premium of 2.5 cents per kWh to support acquisition of wind power by the main utility in the state (Xcel Energy 2001). This program directly funded about 45 MW of wind capacity as of 2001. It also paved the way for wind power development as a competitive resource in the state.[6] The success of green power programs depends on a number of factors, including the price premium, the degree of marketing, and the credibility of the green power provider (Mayer, Blank, and Swezey 1999).

The adoption of a small surcharge on all electricity sales (known as a public benefits or systems benefit charge) is another way to support energy efficiency and renewable energy sources, as well as assist low-income and rural households to obtain adequate energy services, in a more competitive market environment. In the United States, regulators in about 20 states have adopted an electricity surcharge to fund energy efficiency and other public benefit activities (Kushler and Witte 2001). The surcharge is usually a few percent or less of the total electricity price. Some states operate these programs through state agencies, others through distribution utilities or independent program implementors.

Brazil is another country that has adopted a public benefits charge in conjunction with utility privatization and restructuring. In this case, a 1 percent charge on retail service is collected for the purpose of funding R&D and utility energy efficiency programs (Geller 2000). Similar energy efficiency funding schemes were enacted in conjunction with utility restructuring in Denmark, Norway, and the United Kingdom (Nilsson et al. 2001). In Denmark and the United Kingdom, efficiency programs are implemented by regional utilities, whereas in Norway programs are implemented by independent energy efficiency centers.

Utility privatization and restructuring in developing countries often result in tariff increases. When implementing such increases, it is important to protect low-income consumers. Also, private utilities should be regulated and required to expand access to electricity. Ghana, for example, is attempting to increase access to electricity as it restructures and privatizes its power sector. This is being done by requiring that private utilities serve all consumers in urban areas, expand electricity service outside of urban areas, and maintain discounts for low-income households. Also, policy makers are removing barriers to off-grid renewable energy development in rural areas (Edjekumhene, Amadu, and Brew-Hammond 2001).

Creation of rural electric cooperatives is one strategy for expanding electricity supply in rural areas. Rural electric coops, operating as nonprofit organizations with access to low-cost capital, extended electricity service throughout rural America. Establishing rural electric cooperatives as part of power sector reform could help to expand electricity service and use of renewable energy technologies in rural areas of developing countries. In fact, 45 rural electric cooperatives were set up in Bangladesh in the 1980s when the large parastatal utilities experienced severe technical and financial problems. These cooperatives extended electricity supply to 1.6 million rural consumers, improved system reliability, and improved revenue collection (Goldemberg 2000).

Providing long-term concessions to renewable energy developers in off-grid rural areas is another type of market reform that could increase access to electricity and boost renewable energy use. Argentina, for example, is using this approach to promote renewable energy use and electrification of 1.4 million unserved rural households as well as schools and health clinics in northwestern Argentina. About half the cost of renewable energy systems will be recovered from users and about half provided through subsidies from the state and federal government as well as the Global Environmental Facility (Goldemberg 2000).

South Africa is using a similar strategy of granting solar power concessions to public–private partnerships. One is a widely publicized partnership between Shell Solar and ESKOM, the South African national utility. This partnership is using an ESCO approach, whereby solar PV systems are leased to consumers who pay for electricity as it is consumed. But solar PV home systems have failed to take off so far in South Africa due to delays in the introduction of government subsidies and other problems (Hankins 2001).

Market Obligations

Market obligations are a form of regulation that can be used to increase energy efficiency or renewable energy use. Utilities or energy agencies can be required to achieve a specified level of energy savings through energy efficiency programs. This approach has the advantage of establishing savings rather than spending targets and is followed in some European countries. However, the systems benefit charge approach discussed above has been more popular for stimulating energy efficiency in the United States.

Utilities can also be required to supply or purchase a specified amount of electricity from renewable energy sources, expressed as either a fixed amount of capacity or a percentage of total electricity sales. In the United States, 12 states as of 2001 required their utilities to obtain some fraction of their capacity from renewables through what is known as a Renewable Portfolio Standard (RPS). The RPS varies with respect to amount of renewables required and eligible technologies in each state (Table 3-6). Some states limit the amount that can be provided by more established technologies or set aside a portion of the requirement for emerging technologies such as solar electric systems (Nogee et al. 1999, UCS 2001).

Not surprising, renewable energy development has been strongest in the United States where an RPS has been in effect for a number of years. For example, the upper Midwest region, Pacific Northwest, and Texas were the leading

TABLE 3-6

Renewable Portfolio Standards Policies in the United States as of 2001

State	Renewables Requirement	Eligible Technologies
Arizona	0.2% by 2001; 1.1% by 2007	Solar photovoltaic and thermal electric must supply at least half the total
Connecticut	6% in 2001; 13% in 2009	Existing technology (hydro and municipal waste) limited to 7% in 2009; new technologies (wind, new biomass, solar, landfill gas, and fuel cells) must provide 6% by 2009
Hawaii	7% by 2003; 9% by 2010	Wind, solar, biomass, hydro, geothermal, landfill gas, waste, fuel cells, and ocean energy
Iowa	2% by 1999	Wind, biomass, solar, hydro, and waste
Maine	30% by 2000	Biomass, wind, solar, hydro, fuel cells, geothermal, landfill gas, and cogeneration and municipal waste projects under 100 MW
Massachusetts	11% by 2009	Biomass, hydro, municipal waste, wind, solar; existing hydro and municipal waste limited to 7% of total
Minnesota	1% by 2005; 10% by 2015	Wind, biomass, hydro, solar
Nevada	5% by 2003; 15% by 2013	Solar, wind, biomass, and geothermal
New Jersey	5.5% by 2006; 10.5% by 2012	New solar, wind, geothermal, landfill gas, biomass, and fuel cell facilities must provide at least 4% of the total by 2012
Pennsylvania	2%	Wind, solar, biomass, geothermal, and municipal waste
Texas	Approx. 3% by 2009	2000 MW of solar, wind, biomass, hydro, geothermal, and landfill gas
Wisconsin	2.2% by 2011	Solar, wind, geothermal, biomass, fuel cells, and small-scale hydro

NOTES: MW = megawatts.
SOURCES: Berry and Jaccard 2001, DSIRE 2001, Nogee et al. 1999.

regions for new wind power capacity in the United States in recent years.[7] Nearly 900 MW of wind capacity were brought on line in Texas in 2001 alone. Besides diversifying their electricity supplies, these regions are benefiting from new factories that manufacture renewable energy technologies; companies that market, install, and service renewable energy systems; and payments to farmers who lease their land to wind project developers (Singh 2001).

Other countries where an RPS has been adopted or proposed include Denmark, the Netherlands, Italy, Australia, Belgium, and the United Kingdom (Berry and Jaccard 2001, Espey 2001). The United Kingdom, for example, set a 5 percent renewable energy requirement by 2003 and 10.4 percent requirement by 2011 (Cameron, Wilder, and Pugliese 2001). This quantity-based requirement was adopted after a renewable energy obligation implemented through a series of competitive bids during the 1990s met with mixed success (see discussion below).

An RPS provides certainty that renewable energy sources will be implemented, stimulates competition among renewable energy providers, and fosters a least-cost approach to renewable energy development (Espey 2001). Generally, utilities are allowed to achieve their renewable energy goal through installation of renewables facilities and/or purchase of tradable renewable energy credits. The tradable credit scheme helps to minimize the overall cost of compliance. The main issues that must be addressed in designing an RPS are quantity, timing, which resources to include, whether to include a cost cap, and whether to specify minimum (or maximum) amounts for particular types of renewable resources (Nogee et al. 1999).

In 1990, the U.K. government mandated that utilities in England and Wales acquire 1,500 MW of renewable capacity by 2000 in conjunction with electric utility privatization and restructuring (Mitchell 2000). In this case, landfill gas and municipal and industrial wastes were included among eligible resources. This capacity was obtained through a series of competitive request for bids conducted by the central government—a policy known as the Non–Fossil Fuel Obligation (NFFO). Five bids were conducted from 1990 to 1998. This approach led to increasing amounts of capacity proposed and contracted for, and steadily declining prices for the winning bids (Table 3-7). The average price declined from about 7 pence/kWh (US$0.105/kWh) in the early rounds to 2.7 pence/kWh ($0.04/kWh) by the final round in 1998.

There were difficulties in the implementation of the NFFO, however. Only 730 MW of capacity were operating out of 3,270 MW contracted for as of early 2000 (Mitchell 2000). This poor rate of implementation is attributed to complex contracting procedures, some projects developers proposing unrealistic prices to win contracts, environmental and social concerns related to proposed wind power projects, and high transaction costs (Mitchell 2000, Volpi 2000). Also, the focus on minimum cost favored established technologies (e.g., landfill gas and municipal waste projects), large projects, and imported technologies, rather than nurturing a diverse set of renewable technologies and nascent domestic renewable energy industries.

Market obligations can be used to introduce renewable-based fuels into the

TABLE 3-7

Experience with the Non–Fossil Fuel Obligation in the United Kingdom

	Projects Contracted[a]			Average Price	Projects Generating[a,b]	
Round	Year	Number	MW	p/kWh	Number	MW
NFFO1	1990	75	152	7.0	61	145
NFFO2	1991	122	472	7.2	82	174
NFFO3	1994	141	627	4.4	75	254
NFFO4	1997	195	843	3.5	56	133
NFFO5	1998	261	1,177	2.7	17	24
TOTAL		794	3,271		291	730

NOTES: [a]Capacity defined in terms of equivalent baseload power capacity that would produce the same amount of electricity annually.
[b]Projects implemented and generating power as of 2000.
kWh = kilowatt-hour; MW = megawatt.
SOURCE: Mitchell 2000.

transportation sector, thereby cutting petroleum use. For example, gasoline and diesel fuel suppliers could be required to include a specified percentage of renewable fuels such as ethanol and biodiesel in their fuel sales mix, or purchase renewable fuel credits from other fuel suppliers.[8] This type of policy has not been adopted yet, but a renewable fuel content standard for motor vehicle fuels has been proposed in legislation introduced in the U.S. Congress (Urbanchuk 2001).

While there is much less experience with energy savings obligations compared to renewable energy obligations, energy suppliers can be required to achieve specified levels of end-use savings. For example, the national energy regulatory authority in Italy is requiring electricity and gas distributors to achieve predetermined energy savings target during 2002–2006. The savings requirements can be met either through energy efficiency programs or purchase of tradable energy efficiency certificates from energy service companies (Pavan 2002). A similar energy savings obligation has also been imposed on energy suppliers in the United Kingdom. The implementation of a credible and practical project monitoring and evaluation scheme is critical to the success of this type of policy.

Market obligations can also be used to stimulate commercialization of advanced vehicle technologies. California, for example, is requiring that a small fraction of new vehicle sales consist of electric, fuel cell, hybrid gasoline–electric, and other very low emissions or zero-emission vehicles (ZEVs). The ZEV mandate begins at a 10 percent market share in 2003 for all types of cleaner vehicles, with at least a 2 percent market share for "full-function electric vehicles" (CARB

2001). The market share requirements increase over time. Also, the state provides grants of up to $9,000 toward the purchase of electric vehicles to make them more attractive to consumers. This innovative state policy has been modified over time in response to technological and market developments, particularly the slow progress in developing viable battery-powered electric vehicles. But the ZEV program has helped to advance electric-drive technology as well as hybrid gas–electric and very low polluting gasoline-fueled vehicles in the United States (Ogden, Williams, and Larson 2001).

Pollutant emissions cap and trading policies are another type of market obligation that can stimulate clean energy technologies if properly designed. Emissions cap and trading schemes have been successfully implemented to reduce sulfur dioxide emissions and nitrogen oxide emissions in the United States. But tradable emissions allowances were allocated to states based on a percentage of historic emissions, an approach that does not enable energy efficiency and renewable energy options to directly participate in the emissions marketplace. It would be preferable to award emissions allowances to all energy sources including verifiable energy efficiency projects based on the amount of energy produced or saved, or to auction off emissions allowances. Doing so would recognize and reward the emissions benefits of energy efficiency and renewable energy technologies, thereby encouraging their adoption (Shepard 2001, Wooley 2000).

Capacity Building

Capacity building is essential if energy efficiency, renewable energy, and clean fossil fuel technologies are going to make a major contribution to future energy needs. All nations require expertise in a wide range of technical, marketing, managerial, and public policy areas including:

- Technology development, adaptation, assessment, and testing;
- Manufacturing and marketing;
- Sustainable energy business development;
- Deployment and behavioral issues;
- Monitoring and evaluation;
- Training energy managers and end users; and
- Policy development and implementation.

Capacity building is needed at both individual and organizational levels. In particular, there is a need to form and staff public sector agencies, research institutes, and local training, outreach, and service centers. This type of infrastructure was a critical factor in the success of the improved cookstoves, biogas digester, and small wind power programs in China, for example (Martinot et al. 2002, Smith et al. 1993). Capacity building is also needed to staff and train the private companies that will produce, market, install, and service clean energy technologies.

Capacity and institution building in part means forming strong national energy efficiency and renewable energy centers and programs like those successfully operating in Brazil, China, India, Japan, and Eastern Europe (see Box 3-2 and case studies in Chapter 4). These centers carry out the wide range of activities discussed in this chapter, including demonstrations, information dissemination, training, financing, providing financial incentives, and promoting policy reform.

It is critical that such centers and programs cooperate with and support the marketing efforts of the private sector. This means involving the private

BOX 3-2

Energy Efficiency Centers in Central and Eastern Europe

The formerly centrally planned economies of Central and Eastern Europe and the former Soviet Union waste vast quantities of energy. Consequently, many of these nations created national energy efficiency centers during the early 1990s. The centers received foreign assistance initially but later became self-sufficient. The centers assist private firms, demonstrate energy-efficient technologies, provide energy management training, disseminate information, and conduct analysis to support policy reform (Chandler et al. 1999). Among their accomplishments:

- New apartment and commercial buildings in eight Russian regions are now built according to energy standards written by the Russian Center for Energy Efficiency (CENEf). The new building code in Moscow, for example, has led to a 40 percent reduction in energy use in new and renovated buildings (Matrosov, Chao, and Goldstein 2000). CENEf also helped draft the first version of the Russian Federal Law on Energy Conservation, helped develop regional energy efficiency programs, and developed technical and financial plans for district heating system retrofits financed by the World Bank.
- Ukraine invested $6 million to cut staggering energy waste in government buildings and in private factories, thanks to the efforts of

ARENA-ECO, the Ukrainian efficiency center. ARENA-ECO also helped develop a comprehensive national energy efficiency program including energy efficiency standards and an energy efficiency loan fund. And it assisted a number of multinational firms who are now distributing energy-efficient products or services in Ukraine.

- Poland implemented a utility reform law that ensures independent power producers access to the national grid and utility investment in energy efficiency, thanks to legislation drafted by the Polish Foundation for Energy Efficiency (FEWE). FEWE also created an ESCO that has retrofitted major apartment complexes, developed financing mechanisms for energy efficiency projects, and helped to implement a successful efficient lighting program in Poland.
- The Czech Energy Efficiency Center (SEVEn) facilitated ESCO development, drafted portions of a national energy efficiency law, and prepared energy efficiency plans for a number of Czech cities. SEVEn also played a major role in energy efficiency demonstration projects in schools, hospitals, and apartment buildings; developed building energy standards; and contributed to energy price subsidy reform in the Czech Republic.
- The Bulgarian Center for Energy Efficiency (EnEffect) developed a national energy efficiency plan as well as an energy efficiency law. The law authorizes efficiency standards and labeling, creation of a financing fund for energy efficiency projects, financial incentives, and other activities (IEA 1997c). Also, EnEffect created the Municipal Energy Efficiency Network and helped municipalities develop energy efficiency plans.

The energy efficiency centers in Central and Eastern Europe have been successful on a number of levels. First, they facilitated business development and investments for energy efficiency improvement on a sizable scale. Second, they helped to create policy and regulatory structures that are leading to widespread energy efficiency improvements in their countries. Third, they are building the capacity needed to increase energy efficiency over the long run. And finally, they have managed to survive, if not thrive, financially during a period of economic turmoil in their countries (Chandler et al. 1999).

◉ ◉ ◉

sector in program design and implementation. It also means providing training and technical and financial assistance to the private sector, but not directly competing with it. Strong government–private sector collaboration is a key feature of successful wind power deployment in Denmark, improved cookstove and small-scale wind turbine deployment in China, and renewable energy deployment in India (see Chapter 4).

Capacity and institution building should be given greater priority both by national governments and by international assistance agencies. The World Bank and bilateral donors have at times been criticized for failing to build an infrastructure that is capable of marketing and supplying solar energy and other renewable energy devices over the long run once a particular "development project" ends (Martinot et al. 2002, Mulugetta, Nhete, and Jackson 2000).

Planning Techniques

Many countries and regions have developed energy plans for increasing energy efficiency and renewable energy use, providing energy services in a cost-effective manner, and reducing the adverse environmental impacts due to energy production and use. In recent years, many of these plans aim to reduce CO_2 emissions. To be successful, energy plans should contain achievable goals, measures and actions for achieving the goals, and monitoring and evaluation procedures.

The energy plan adopted and implemented in Upper Austria is an example of successful energy planning. Upper Austria is a highly industrialized region with 1.4 million inhabitants. It developed an energy plan in 1991 that promoted greater energy efficiency and renewable energy use through information and education efforts, grants and loans, financing, R&D, and some regulatory measures. Specific targets to be met by 2000 were established in each sector. The plan appears to have been well implemented, and most of the goals are expected to be achieved. By 1996, Upper Austria achieved nearly a 12 percent reduction in CO_2 emissions (Egger and Dell 1999).

Integrated resource planning (IRP) is a process whereby a planning authority identifies the mix of supply- and demand-side resources that provide energy service needs at the lowest cost (NARUC 1988, Swisher, Jannuzzi, and Redlinger 1997). The objective is to provide services such as heat, light, refrigeration, and motive power—not energy per se—as cost-effectively as possible. This leads to consideration of energy efficiency options as resources on par with supply-side options. IRP has been successfully applied in some parts of the United States. For example, IRP is practiced in the Northwest region, resulting in large-scale energy efficiency programs that have achieved significant energy savings (Box 3-3). When carrying out IRP, it is possible to account for environmental costs not already included in market prices, thereby selecting

BOX 3-3

Regional Energy Planning in the Pacific Northwest

The Northwest Power and Conservation Act of 1980, a law adopted by the U.S. federal government, established integrated resource planning in the northwestern states of Oregon, Washington, Idaho, and Montana. The act includes four long-term goals: (1) achieve cost-effective conservation, (2) encourage development of renewable energy resources, (3) establish a regional power planning process, and (4) assure the region of an adequate and economical power supply while protecting the environment.

The act created the regional Power Planning Council that prepares conservation and electric power plans at least once every five years. The plans are based on projected energy supply costs for different types of resources, potential energy conservation and its costs, and quantification of environmental costs and benefits. These plans resulted in over $1 billion in energy efficiency investments by utilities in the Pacific Northwest from 1980 to 1997, resulting in about 11 TWh (6 percent) electricity savings as of 1997 at an average cost of 2 to 2.5 cents per kWh saved. Energy efficiency improvement cost about half as much as new generating resources available at the time, and provided about 21 percent of the new electricity "resources" acquired in the region during the period from 1991 to 1997 .

The 1998 regional power plan identified additional cost-effective efficiency measures that would provide 13 TWh per year of electricity savings within 20 years. Achieving this goal would save the region's consumers an estimated $2.3 billion on their future energy bills and would cut electricity use in the region by about 7 percent at the end of the period (in addition to previous savings). When the region was confronted with low hydroelectric output and other problems in 2001, states and utilities in the region stepped up their energy efficiency efforts to prevent power shortages. The Power Planning Council proposed a short-term goal of saving the equivalent output of a new 300 MW power plant over three years.

The 1980 act also directs the Power Planning Council to protect and enhance fish and wildlife, and their habitat, in the region. Some 79 hydroelectric plants and hundreds of other irrigation dams have been built over the past 65 years throughout the Columbia River Basin in the northwestern United States. This has dramatically reduced the number of salmon and steelhead trout, vital natural resources living and breeding in the region's rivers.

The Power Planning Council has developed programs for fish and wildlife protection and recovery including reducing hydroelectric capacity and increasing water flows during critical periods. The Bonneville Power Administration, the primary power-generating authority in the region, spent about $3.4 billion on fish and wildlife protection and recovery (including the lost revenue from altering dam operations) from 1982 to 1999. While salmon and steelhead recovery has proven to be very difficult, the region is committed to balancing energy production, conservation, and natural resource protection and recovery. Integrated resource planning plays a central role in this commitment.

SOURCES: NPPC 1998, 2001, Wilkinson 1992.

◉ ◉ ◉

resources based on a more complete cost accounting than that occurring in the marketplace.

An IRP perspective can be valuable in developing countries where both modern energy services and investment capital are limited. IRP can direct energy authorities and utilities to the largest and most cost-effective energy savings opportunities. IRP can also identify renewable energy technologies as the least-cost approach to rural electrification. IRP studies have helped to increase support for end-use energy efficiency efforts in Brazil, India, and Sri Lanka (Geller 1991, Padmanabhan 1999). IRP is also getting under way in Ghana and South Africa.

Integrated land use and transport planning is a strategy that can lead to more efficient urban design and greater reliance on public transit systems, compared to the car-intensive urban sprawl found in many cities. Table 3-8 compares the density, transport mode split, transport energy consumption, and transport cost in six major cities. It is clear from this comparison that denser cities like Hong Kong, Singapore, and Munich tend to be much less car-dependent and less energy-intensive than sprawling cities such as Houston. Denser cities with better public transport systems also tend to be less costly and have fewer transport-related fatalities than car-intensive and sprawling cities (Vivier and Mezghani 2001).

TABLE 3-8

Comparison of Urban Density and Transport Mode Choice, Energy Intensity, and Cost in Six Cities

City	Density (inhabitants/ha)	Modal Split[a]		Energy Intensity (GJ/inhabitant)	Transport Cost (% of economic output)
		Cars	Others		
Houston	9	95.5	4.5	86	14.0
New York	18	75	25	43	9.4
Paris	48	44	56	15.5	6.8
Munich	56	40	60	17.5	5.8
Singapore	94	53	47	12	4.7
Hong Kong	320	18	82	6.5	5.0

NOTES: [a]Modal split is the percentage of trips by cars on the one hand and public transport, walking, and bicycling on the other. GJ = gigajoules, ha = hectare.
SOURCE: Vivier and Mezghani 2001.

Curitiba, Brazil, exemplifies the potential for providing efficient transportation services through careful land use and transport planning. Curitiba is a rapidly growing city with a population of over 2 million. Starting in the 1970s, Curitiba implemented urban planning and a sophisticated public transportation infrastructure aimed at reducing automobile use and urban sprawl. The city built an extensive bus system including express bus lanes and efficient transfer stations, and development was concentrated around mass transit corridors. As a result, about 75 percent of commuters use buses, fuel use per capita is about one-fourth lower than in comparable Brazilian cities, and ambient air pollution is relatively low (Hawken, Lovins, and Lovins 1999, Rabinovitch and Leitman 1996).

Freiburg, Germany, offers another example of integrated land use and transport planning, combined with other policies to promote more sustainable personal transportation. Freiburg has maintained a compact land development pattern, limited urban sprawl, closed its downtown to private cars, increased parking fees, introduced an inexpensive public transit pass, and improved its public transportation network. As a result, ridership on public transport more than doubled between 1983 and 1995, and close to 60 percent of trips were by public transport, bicycle, or foot as of 1992 (FitzRoy and Smith 1998).

Stockholm, Sweden, has developed its extensive public transit system in coordination with careful land use planning. New communities were located close to rail links as the city grew, rail fares were kept relatively low, and parking fees in the city center were increased. As a result, about half of Stockholm's workers commute by bus or rail, and car use per capita declined from 1980 to 1990

(Tellus Institute 1999). This has been achieved in an affluent city with a high degree of car ownership.

Summary

This chapter reviewed the different types of policies that can be used to overcome the barriers to greater energy efficiency and renewable energy use. As shown in Table 3-9, each type of policy addresses particular barriers.

The mounting experience with the implementation of these policies in both industrialized and developing nations provides a number of overarching lessons, which are summarized below by type of policy.

Research, development, and demonstration. Government-funded RD&D can be an important policy for advancing energy efficiency and renewable energy

TABLE 3-9

Barriers and Policies for Overcoming Them

Barrier	Policies
Limited supply infrastructure	Research, development, and demonstration Regulations Procurement Market obligations Capacity building
Quality problems	Research, development, and demonstration Regulations Information dissemination and training Capacity building
Insufficient information and training	Information dissemination and training Capacity building
Misplaced incentives	Financial incentives Regulations Planning techniques
Lack of money or financing	Financing Financial incentives
Purchasing procedures	Financing Financial incentives Information dissemination and training Voluntary agreements Regulations Procurement
Pricing and tax barriers	Financial incentives Pricing Regulations
Regulatory and utility barriers	Pricing Regulations Market reforms Market obligations Planning techniques
Political obstacles	Regulations Market obligations Planning techniques

technologies. RD&D tends to be more effective when it involves collaboration between research institutes and the private sector; when it is coordinated with other policies such as financial incentives, market reserves, or regulations; and when it focuses on a broad range of designs. RD&D on clean energy technologies merits expansion. In addition, greater international collaboration in RD&D on clean energy technologies, including collaboration between industrialized and developing nations, could provide a number of benefits including cost and risk sharing, more rapid learning, and faster deployment of clean energy technologies worldwide.

Financing. Financing can help increase the adoption of energy efficiency and renewable energy technologies, especially in developing countries. Financing schemes should be designed to support energy efficiency and renewable energy businesses such as ESCOs or solar energy entrepreneurs. Also, financing schemes should be designed to minimize transaction costs, support high-quality products, and work through established financing channels such as commercial banks or rural credit cooperatives where possible. Financing without subsidies can be a viable long-term policy, but lower income groups in all likelihood will need subsidies to afford clean energy devices. Financing is most effective in combination with other policies such as financial incentives and the creation of a marketing, delivery, and service infrastructure.

Financial incentives. Financial incentives provided by governments or utilities can be an effective tool for stimulating the adoption of energy efficiency and renewable energy technologies. Incentives should reward energy savings or renewable energy output rather than reward investment; be sustained over many years to build up markets and drive down costs; and phase out gradually as costs drop and other barriers are removed. Financial incentives can also be used to stimulate commercialization and initial markets for innovative technologies. In designing incentive programs for developing countries, it is important to support rather than hinder local manufacturing and marketing of energy efficiency and renewable energy measures.

Pricing. Taxing fossil fuels based on their social and environmental costs can increase energy efficiency, reduce energy consumption, and encourage introduction of renewable energy sources. Taxing the carbon content of fossil fuels can be an effective policy if the tax is large enough and a portion of the revenues is used for incentives for energy efficiency and renewable technologies. It is also important to reduce or eliminate subsidies on conventional energy sources in countries where they still exist, but this should be done in ways that do not harm low-income consumers. Paying avoided costs or retail prices for renewable energy or cogenerated power supplied to the electric grid can support the adoption of these technologies. Finally, differential taxes based on the relative energy efficiency of different products or activities is a promising but underutilized energy policy.

Voluntary agreements. Voluntary agreements between government and the private sector can be an effective strategy for increasing energy efficiency. Voluntary agreements tend to work best when industries fear they will face taxes or regulations if meaningful energy efficiency targets are not established and met, when thorough monitoring and evaluation take place, and when companies that achieve significant results are recognized. In addition, states and municipalities are starting to make voluntary commitments to increase energy efficiency or renewable energy use. These commitments work best if adopted along with specific policies that will facilitate achieving the targets, such as financing, incentives, or regulations.

Regulations. Appliance, vehicle, and building efficiency standards can stimulate energy efficiency improvements on a large scale. Experience around the world demonstrates that it is important to update these standards periodically to maintain their influence. Also, it is important to monitor compliance and rigorously enforce efficiency standards, especially energy codes for new buildings. In addition, reasonable performance standards can improve the quality of renewable energy technologies such as household PV systems.

Information dissemination and training. Information and training can help address some of the barriers limiting the adoption of energy efficiency measures and renewable energy sources. Efficiency labeling can be useful if it clearly identifies high-efficiency products for consumers. Training can be valuable for ensuring that energy efficiency and renewable energy technologies are installed and used properly. And information dissemination and promotion can increase the awareness and adoption of renewable energy measures. But information dissemination and training tend to be more effective when they are targeted to decision makers at the time of appliance purchase or building construction, or when combined with other policies such as financing, incentives, voluntary agreements, or regulations.

Procurement. Government-led procurement can help to commercialize and build markets for innovative energy efficiency and renewable energy technologies. Government-led procurement has been used mainly for energy efficiency measures so far, but governments are starting to purchase significant amounts of renewable energy technologies and green power. Government procurement should be carried out in ways that support the development of viable markets for these technologies over the long run (e.g., by working with vendors and phasing out subsidies over time).

Market reforms. Utility sector privatization and increased competition can lead to efficiency improvements and emissions reductions in the supply of electricity. But utility sector restructuring is unlikely to advance end-use energy efficiency, renewable energy development, and increased access to electricity supply unless specific policies are adopted to address these needs. Such policies include renewable energy obligations, encouraging marketing of green power, rate discounts for low-income households, and small surcharges on retail electricity service to fund energy efficiency and other clean energy activities. In addition, rural electric cooperatives and renewable energy concessions are two promising market strategies for increasing electrification of rural households in an environmentally sound manner.

Market obligations. Market obligations can result in substantial acquisition of renewable energy sources or energy efficiency measures. In particular, renewable portfolio standards are expanding renewable energy implementation in the electricity sector in the United States and other countries. Renewable energy acquisition is done on a competitive basis, thereby driving down the cost of renewable electricity sources. A few European nations have begun to impose energy savings obligations on their energy suppliers. This policy deserves wide consideration, especially if the pioneering European efforts are successful. Market obligations can also be used to stimulate the development and commercialization of innovative vehicle technologies or renewable fuels. Likewise, emissions cap and trading schemes can help to stimulate energy efficiency improvements and renewable energy development if these policies are carefully designed.

Capacity building. Capacity building is critical for implementing clean energy technologies on a large scale. All countries need energy efficiency and renewable energy centers and programs at the national, state, and local levels. Capacity building is also needed to staff and train private companies that manufacture, market, and service these technologies. Capacity building should be given greater priority in energy assistance projects for developing and transition countries.

Planning techniques. Careful energy and transportation planning can help nations, regions, and cities move toward a more sustainable energy future. Integrated resource planning helps to define the optimal set of demand-side and supply-side investments for meeting future energy service needs. It can include environmental considerations and it usually leads to increased investment in energy efficiency measures. Integrated land use and transport planning helps to expand public transportation systems, locate new housing and commercial development near public transport facilities, and reduce urban sprawl. Doing so can cut fuel use and provide other benefits.

The review of individual policies frequently illustrated that policies work best if implemented in combination. This finding is understandable given the variety of barriers that exist in most countries and markets. Chapter 4 presents 10 examples from around the world where various policies have been integrated into successful market transformation strategies.

Notes

1. Electricity suppliers are required to carry out energy efficiency programs in the United Kingdom.
2. Per unit of energy, the carbon content of coal is about 80 percent greater than natural gas. Thus, a carbon tax disproportionately affects the price of coal and provides an advantage to low-carbon fuels and renewables.
3. It is worth noting that there were flaws in the PURPA requirements such as relatively low efficiency thresholds and no allowance to modify contracts as avoided costs changed over time.
4. These programs were modified in the late 1990s to focus on whole building energy performance. Commercial buildings are now given the Energy Star® designation if they are among the "top 25 percent" of a particular building type with respect to energy efficiency.
5. Green power has a relatively high renewable energy content, in some cases 100 percent renewables.
6. Wind power was able to compete favorably with fossil fuel technologies in part because it receives a 1.7 cent per kWh production tax incentive from the federal government.
7. The Pacific Northwest states have not adopted the RPS policy, but they have enacted other policies supporting renewable energy development, including requiring electricity providers to offer customers a green power option.
8. A renewable fuel standard would work in the same way as an RPS for electricity.

Market Transformation

The market transformation approach has been developed, tested, and evaluated over the past 10 years in the energy efficiency and renewable energy fields. A market transformation strategy attempts to remove barriers to achieve a permanent shift in the market. The market effects can be defined as sales levels or market shares for energy-efficient products, energy-saving practices, or renewable energy technologies. Intermediate effects such as increased availability of the desired technologies, or meeting predetermined performance or cost goals, are important as well. The ultimate objective is to make the energy efficiency or renewable energy technology or practice the norm through a set of coordinated market interventions (Eto, Prahl and Schlegel 1996, Geller and Nadel 1994).

The market transformation approach is consistent with the concept of technological learning and experience curves. This concept shows that the cost of producing a technology decreases as manufacturers' experience accumulates (IEA 2000e, McDonald and Schrattenholzer 2001). Policy interventions help to remove barriers inhibiting adoption and thereby increase sales of new technologies, which in turn

TABLE 4-1

Clean Energy Technology Learning Rates

Technology	Country/Region	Time Period	Learning Rate (%)
Solar photovoltaic modules	U.S., Japan	1981–95	20
Solar photovoltaic systems	Europe	1985–95	35
Wind power	Denmark	1982–97	8
Wind power	California	1980–94	18
Ethanol fuel	Brazil	1979–95	20
Compact fluorescent lamps	U.S.	1992–98	16
Gas turbine, combined-cycle power plants	World	1991–97	26

NOTES: The learning rate is the average reduction in cost for each doubling in cumulative output.
SOURCES: IEA 2000e, McDonald and Schrattenholzer 2001, Nakicenovic, Grubler, and McDonald 1998.

results in unit cost reductions. This creates a positive feedback loop that can enable rapid market growth.

Some activities such as research, development, and demonstration (RD&D) can help to increase the "learning rate" (i.e., the percentage reduction in cost for each doubling in cumulative output). These learning rates tend to be highest during the "shakeout stage" when the number of low-cost producers expands and prices rapidly fall, or when a technological breakthrough is made (IEA 2000e). Table 4-1 presents the empirical learning rate for a number of the energy efficiency and renewable energy technologies. The learning rate averages about 20 percent for these five technologies.

This chapter presents 10 case studies that demonstrate the deployment of energy efficiency or renewable energy technologies on a large scale. Some efforts have already resulted in market transformation and self-sustaining adoption of efficiency improvements or renewable technologies. Other efforts are still in midstream, although they have clearly achieved significant market impacts.

The case studies are organized by topic and scale, starting with energy efficiency efforts at the national level, followed by efficiency efforts at the sectoral or state level, and then renewable energy efforts. The final case study is an example of shifting to natural gas–based electricity generation at the national level.

China: National Energy Efficiency Program

Inefficient energy use was rampant in China during the 1960s and '70s. In the early 1980s, China began a national energy efficiency program to reduce investment requirements for energy supply expansion and prevent energy from becoming a brake on economic growth. The program focused mainly on industrial energy use and included regulations, a monitoring system for industrial facilities, an energy efficiency financing scheme, and support for RD&D (Table 4-2). Energy management units were established in major energy-consuming enterprises. By 1983, over 10 percent of China's energy-related investments were in energy efficiency. As a result, energy demand growth from 1981 to 1986 fell to half that of economic growth (Sinton, Levine, and Qingyi 1998).

Additional policies implemented in the late 1980s included the creation of over 200 provincial and local energy conservation service centers, energy intensity standards for boilers and kilns, more favorable terms for energy efficiency loans, and allowing firms to retain the financial benefits of energy savings projects. Also, steps were taken to increase the average efficiency of power production (e.g., through building larger, more efficient power plants).

This comprehensive energy efficiency program was extremely successful. As shown in Figure 4-1, China's energy intensity fell by more than 50 percent between 1980 and 1997 (Zhang 1999).[1] Total energy use and carbon emissions more than doubled during this period due to population growth, high economic growth, and increasing amenities and standards of living. But energy consumption and emissions would have grown much faster had these energy efficiency gains not occurred.

Part of this reduction in national energy intensity was due to structural shifts such as shifting production from smaller scale to larger industrial plants. But

TABLE 4-2

Policies Enacted to Stimulate Energy Efficiency in China

- National energy efficiency investment fund with attractive loan rates
- Energy management units and a consumption monitoring system for large industrial facilities
- Regulations related to energy supply quotas, purchase of more efficient equipment, and disposal of outdated equipment
- Applied energy research, development, and demonstration
- Organizations established within the State Planning Commission to coordinate and oversee energy efficiency policy and programs
- Promotion of industrial cogeneration
- Energy intensity standards for industrial boilers, kilns, and other products
- Creation of energy conservation centers to provide training and technical assistance at the provincial and local levels
- Financial incentives offered to firms based on the level of energy savings.

SOURCE: Sinton, Levine, and Qingyi 1998.

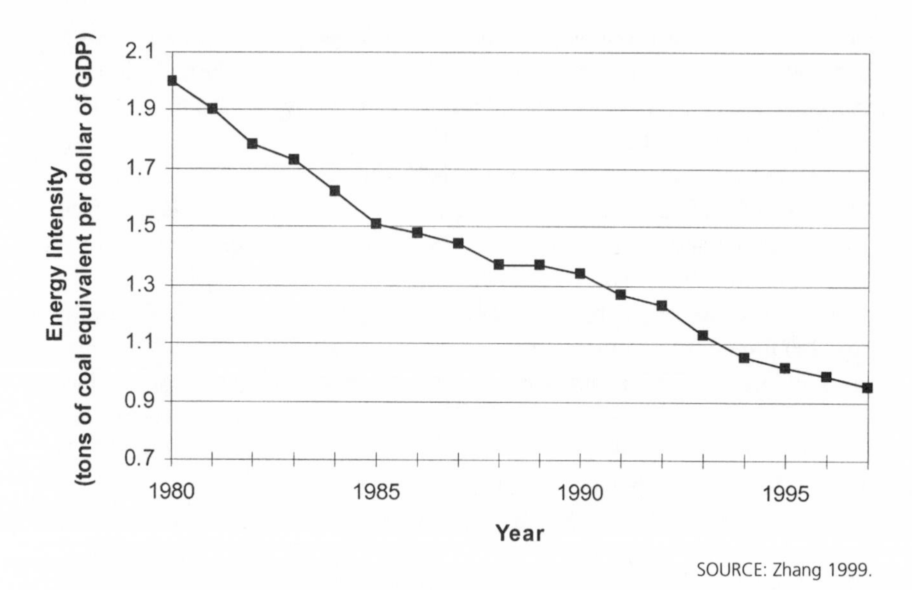

FIGURE 4-1: Overall energy intensity trends in China.

most of the reduction is attributed to technical efficiency improvements (Sinton, Levine, and Qingyi 1998). For example, the use of inefficient open hearth furnaces for steelmaking fell from about 50 percent of capacity in 1970 to less than 15 percent in 1995, contributing to a 20 percent decline in the average energy intensity of steel production from 1980 to 1990 (Phylipsen et al. 1999). Likewise, the use of the inefficient wet process for cement production was phased out by 1990, contributing to an 18 percent decline in the average energy intensity of cement production from 1980 to 1994 (Phylipsen et al. 1999).

China is continuing to enact policies that will stimulate energy efficiency improvements, including cutting energy price subsidies. The price of oil products was increased by a factor of 4 and coal by a factor of 3 between 1990 and 1997 (Sinton and Fridley 2000). China initiated a national lighting efficiency program in 1996 that has significantly increased the production, awareness, and use of energy-efficient lighting products such as compact fluorescent lamps (CFLs) and higher-efficiency fluorescent lamps and ballasts (Nadel et al. 1999). In 1997, an Energy Conservation Law was enacted that calls for stronger energy management efforts in industries, labeling and efficiency standards for mass-produced products, building energy codes, and further financial incentives (Sinton, Levine, and Qingyi 1998). The law provides the basis for maintaining energy efficiency improvements in China, but regulations for implementing the law still need to be developed. In addition, China has undertaken specific projects to improve the efficiency of its refrigerators, boilers, and motors (Martinot and McDoom 2000, Nadel et al. 2001).

As already noted, increasing power generation efficiency was one of the factors contributing to China's dramatic decline in energy intensity over the past 20 years. In rapidly growing Guandong Province, the average efficiency of coal-fired power plants increased significantly between 1990 and 1998. However, there are still many inefficient smaller power plants in operation (Zhang, May, and Heller 2001). Government policies such as more complex and time-consuming approval processes for larger plants and cost-plus electricity pricing have resulted in generation efficiency increasing more slowly than it could have.

Official figures show that overall energy consumption dropped in China in 1998 and 1999 in spite of continued economic growth, with total energy use in 1999 reported to be about 15 percent below that in 1996–97. This steep decline in absolute energy use is attributed to improvements in coal quality (which increases the efficiency of coal use), continued closing of smaller, less efficient power plants, responses to energy price increases, industrial restructuring such as a shift toward larger manufacturing facilities, and a continued shift to cleaner, more efficient fuels in the residential sector (Sinton and Fridley 2000). Although part of this sharp drop in official energy use may be due to inaccurate data, it is clear that China's energy intensity has continued to rapidly fall. Moreover, it is estimated that absolute carbon dioxide emissions from fossil fuel combustion in China declined nearly 9 percent from 1996 to 2000 (Streets et al. 2001).

In summary, China greatly reduced its energy intensity over the past 20 years. Some of this was due to structural change and modernization, but much of it was technological in nature. A comprehensive set of policies, including deep cuts in price subsidies, efficiency regulations, financial incentives, RD&D efforts and extensive education, training, and technology transfer initiatives, played an important role in this achievement. Furthermore, China continues to demonstrate a strong commitment to increasing energy efficiency and reducing energy waste.

United States: Appliance and Vehicle Efficiency Improvements

The energy efficiency of domestic appliances such as refrigerators, freezers, air conditioners, clothes washers, and heating equipment dramatically improved in the United States during the past 30 years. For example, the average electricity use of new U.S. refrigerators declined from 1725 kWh per year in 1972 to about 500 kWh per year in 2001 (Fig. 4-2). This reduction in electricity use was achieved

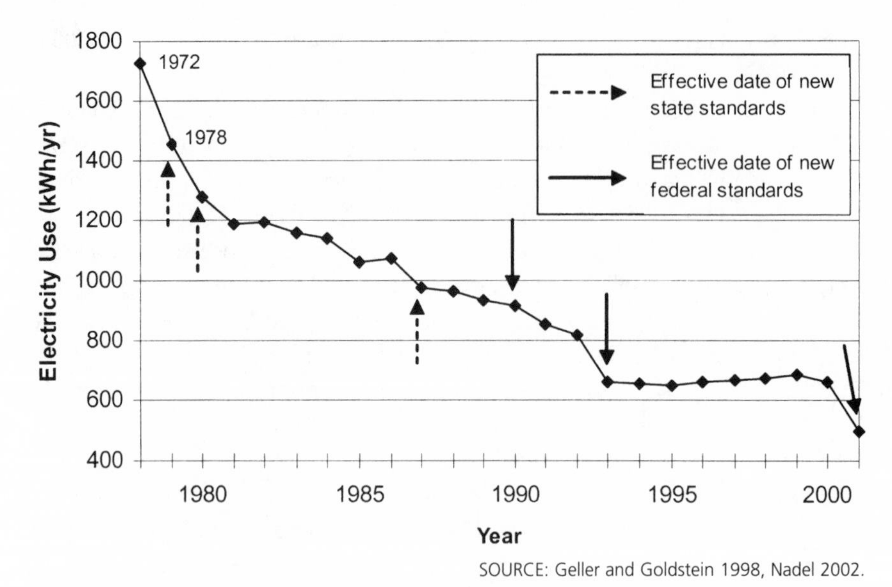

SOURCE: Geller and Goldstein 1998, Nadel 2002.

FIGURE 4-2: Average electricity consumption of new U.S. refrigerators.

even while average refrigerator size increased and the number of features grew (Geller and Goldstein 1998). Also, the average efficiency of new central air conditioners increased about 54 percent and the efficiency of new window air conditioners about 52 percent between the early 1970s and late 1990s.

Government-funded RD&D helped manufacturers develop more efficient appliances and appliance components (NAS 2001a). Labeling of efficient products and utility incentive programs helped to build awareness and stimulate consumer demand for efficient products (Geller and Nadel 1994). But the adoption of mandatory efficiency standards—first at the state level and later at the national level—greatly contributed to the efficiency improvements that occurred in all new products. National efficiency standards were established in laws adopted in 1987, 1988, and 1992 (Geller 1997, Nadel 2002).

These laws include initial energy performance standards, but also direct the Department of Energy to periodically review the standards and adopt more stringent standards if this is found to be technically and economically feasible. The national standards on refrigerators and freezers were significantly strengthened twice (Fig. 4-2). New standards on fluorescent lighting ballasts were adopted in 2000, followed by new standards on water heaters, clothes washers, and central air conditioners. In addition, the U.S. Department of Energy was in the process of adopting revised standards on furnaces, boilers, and commercial air conditioning equipment as of 2002 (Nadel 2002).

It is estimated that appliance and equipment efficiency standards already enacted reduced overall U.S. electricity use by 88 TWh (2.5 percent) in 2000 (Table 4-3). The savings are expected to grow to 253 TWh (6.5 percent) in 2010 due to some new standards taking effect and the stock of existing appliances turning over. The standards usually lead to a slight increase in the first cost of new products, but the operating savings exceed any increase in first cost, often by a wide margin (Nadel 2002). It is estimated that appliance and equipment efficiency standards already enacted will save U.S. consumers over $186 billion net between 1990 and 2030 (Geller, Kubo, and Nadel 2001). In addition, it is estimated that the standards will reduce U.S. carbon emissions by about 61 million metric tons in 2010.

U.S. appliance manufacturers supported the enactment of national minimum efficiency standards as an alternative to a growing patchwork of state standards during the 1980s. Manufacturers and energy efficiency advocates negotiated

TABLE 4-3

Energy Savings and Carbon Emission Reductions from U.S. Appliance Efficiency Standards

	Year Enacted	Electricity Savings (TWh/yr)		Primary Energy Savings (quads/year)		Avoided Carbon Emissions (MMT/yr)	
Products		2000	2010	2000	2010	2000	2010
Major residential appliances	1987	8	41	0.21	0.55	3.7	10.0
Fluorescent lamp ballasts	1988	18	23	0.21	0.27	4.4	5.0
Residential appliance updates	1989/91	20	37	0.23	0.43	4.8	8.1
Lamps, motors, and commercial heating and air conditioning equipment	1992	42	110	0.59	1.51	11.8	27.5
Refrigerator and room air conditioner updates	1997	—	15	—	0.14	—	3.2
Ballast update	2000	—	6	—	0.06	—	1.3
Clothes washer, water heater updates	2001	—	10	—	0.19	—	3.6
Central air conditioner and heat pump updates	2001	—	11	—	0.11	—	2.3
TOTAL	—	88	253	1.2	3.3	25	61

NOTES: TWh = terawatt-hour; quad = quadrillion Btu; MMT = million metric tons.
SOURCE: Geller, Kubo, Nadel 2001.

consensus standards on a number of occasions. For example, appliance manufacturers and efficiency advocates agreed on new standards for clothes washers that will take effect in two stages—in 2004 and 2007. When given adequate lead time to develop and produce complying products, manufacturers have come to accept and in some cases support stringent efficiency standards. The standards enable manufacturers to produce and sell higher-quality, higher-value products, and the standards force all manufacturers to meet the same performance thresholds.

In some cases, utility incentive programs helped lay the groundwork for new, more stringent efficiency standards. This was the case when 24 utilities jointly sponsored a Super Efficient Refrigerator Program, providing $30 million in incentives for high-efficiency models. Appliance manufacturers competed to develop the most efficient products, and incentives were subsequently paid to the manufacturer of the winning products based on sales achieved. The winning products also served in part as the basis for future national efficiency standards (Nadel 2002). Also, regional and utility incentive programs helped build early markets for resource-efficient, energy- and water-saving clothes washers (Suozzo and Thorne 2001). These washers will become the norm in the United States when new clothes washer standards take effect.

The United States adopted energy efficiency standards on cars and light trucks, known as Corporate Average Fuel Economy (CAFE) standards, in 1975. These standards were largely responsible for the near doubling in the average fuel economy of new cars and the more than 50 percent increase in light truck fuel economy from 1975 to 1985 (Fig. 4-3). Had these improvements not occurred, the U.S. car and light truck fleet would have consumed an additional 3 million barrels of gasoline per day and emitted an additional 150 million metric tons of carbon as of 1995 (Greene 1999). The gasoline savings has also meant lower oil imports and consequently lower trade deficits in the United States.

The CAFE standards were met mainly through technical changes such as engine efficiency improvements, weight reduction, and drivetrain improvements. In general, there were no negative side effects, although there is some debate about the whether the standards adversely affected vehicle safety (Greene 1999, NAS 2001b). It is clear that cars and light trucks were made safer and less polluting at the same time that they were made more fuel efficient (e.g., the number of occupant deaths per million vehicles fell over 45 percent from 1979 to 1999) (NAS 2001b). But it is possible (although not conclusive) that safety would be slightly better today had vehicle size and weight not fallen during the 1970s and 1980s.

In addition to the CAFE standards, a tax on inefficient "gas guzzlers" contributed to the rise in vehicle fuel economy in the United States during the late

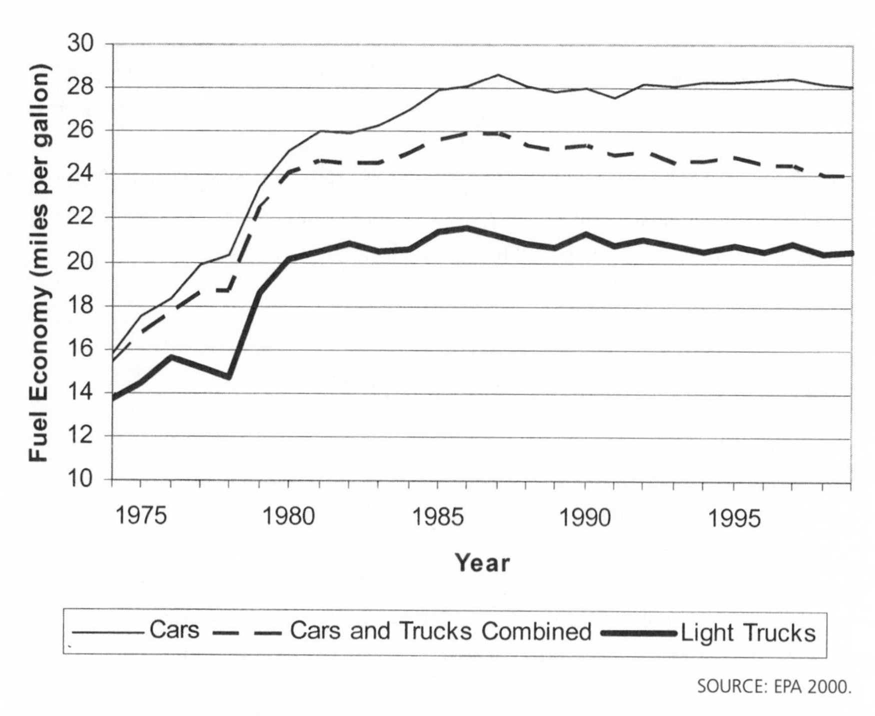

SOURCE: EPA 2000.

FIGURE 4-3: New car and light truck fuel economy in the United States.

1970s and 1980s. Auto manufacturers tended to increase the fuel economy of their least efficient models to minimize gas guzzler tax payments (Geller and Nadel 1994). Today the tax applies to relatively few new cars. Inefficient sport utility vehicles and other types of light trucks are not covered by the tax.

The CAFE standards for cars reached their maximum level in 1985; small increases in the standards for light trucks have been adopted since then. In fact the average fuel economy of new cars and light trucks combined actually declined from a high of about 26 miles per gallon in 1987 to about 24 miles per gallon in 2000 because of fuel economy stagnating and the shift from cars toward less efficient sport utility vehicles, pickup trucks, and minivans (see Fig. 4-3). As a result, U.S. gasoline consumption increased 22 percent from 1985 to 2000 (EIA 2001a). Strengthening the standards is technically and economically feasible (Friedman et al. 2001, NAS 2001b) and has been recommended since the early 1980s (Ross and Williams 1981). However, vehicle manufacturers have vigorously opposed raising the CAFE standards and have prevented such action.

In summary, the energy efficiency of new appliances and passenger vehicles increased dramatically in the United States beginning in the 1970s. While the efficiency of appliances steadily improved over the past 25 years, the efficiency of new vehicles leveled off in the mid-1980s and has slightly declined since then. The adoption of mandatory efficiency standards was the main policy that caused these efficiency improvements. The U.S. experience demonstrates the importance of updating efficiency standards periodically as technologies evolve and older standards become outdated. It also shows that complementing efficiency standards with RD&D, consumer education, and financial incentives can facilitate the adoption and implementation of stringent efficiency standards.

Brazil: More Efficient Electricity Use

The government of Brazil established a national electricity conservation program known as PROCEL in 1985. PROCEL is based at Eletrobras, Brazil's national utility holding and coordinating company. PROCEL promotes end-use electricity conservation as well as transmission and distribution system loss reduction. PROCEL operates by funding or cofunding a wide range of energy efficiency projects carried out by state and local utilities, state agencies, private companies, municipalities, universities, and research institutes. These projects pertain to:

- Research, development, and demonstration;
- Education and training;
- Testing, labeling, and standards;
- Marketing and promotion;
- Private sector support (e.g., ESCO support);
- Utility demand-side management programs; and
- Direct implementation of efficiency measures.

PROCEL also helps utilities obtain low-interest loans for major energy efficiency implementation projects from a loan fund within the electric sector.

PROCEL's annual budget, including grants and low-interest loans but excluding staff salaries and overhead, reached about $50 million in 1998. Some of the key initiatives PROCEL has undertaken are shown in Table 4-4. Many of these initiatives, such as testing and labeling of appliances, motors, and lamps, were conducted in collaboration with the manufacturers of these products. Others were conducted through state and local utilities.

PROCEL's cumulative efforts reduced electricity consumption and supply-side losses in Brazil by about 5.3 TWh per year as of 1998, equivalent to 1.8 percent of electricity consumption that year (Geller et al. 1999). The electricity savings realized in 1998 were about three times that in 1995 (Fig. 4-4). Electricity savings resulted mainly from: (1) increasing the energy efficiency of refrigerators

TABLE 4-4

Major Actions to Improve Equipment Efficiency in Brazil

Refrigerators and freezers
- National efficiency testing and labeling program
- Voluntary energy efficiency targets specifying the maximum electricity use of different types of products as a function of volume
- Recognition and awards for the top-rated models
- Pilot rebate programs for the top-rated models
- Revisions of the test procedure and label, and a new voluntary agreement for efficiency improvements

Lighting
- Replacement of over 1 million inefficient incandescent or "self-ballasted" type street lights with either high pressure sodium lamps or mercury vapor lamps
- Demonstrations, specific utility incentive programs, energy audits, labeling, and television advertisements to promote use of compact flourescent lamps
- R&D and demonstration programs, audits, and educational activities to promote use of T8 lamps, electronic ballasts, and specular reflectors in fluorescent lighting
- Minimum efficiency standards for electromagnetic ballasts

Motors and motor systems
- Technical support to improve the thermal treatment of the carbon steel used in most motor cores
- Establishing minimum efficiency levels for high-efficiency motors sold in Brazil
- Efficiency testing and labeling program for all three-phase induction motors
- Recognition and awards for the most efficient standard motors offered in the marketplace

SOURCE: Geller 2000.

and freezers through testing, labeling, and voluntary agreements with manufacturers; (2) increasing the efficiency of motors through testing, labeling, and R&D projects; (3) increasing the market for energy-efficient lighting technologies such as high-pressure sodium lamps and compact fluorescent lamps; (4) reducing electricity waste in industry through audits, workshops, and information dissemination; and (5) installing meters in previously unmetered households (Geller et al. 1998).

The electricity savings as of 1998 enabled utilities in Brazil to avoid constructing about 1,560 MW of new generating capacity, meaning around $3.1 billion of avoided investments in new power plants and associated transmission and distribution infrastructure. In contrast, PROCEL and its utility partners spent about $260 million on energy efficiency and power supply improvement projects between 1986 and 1998. Thus, from the utility sector perspective, PROCEL has achieved an overall benefit–cost ratio of around 12:1.

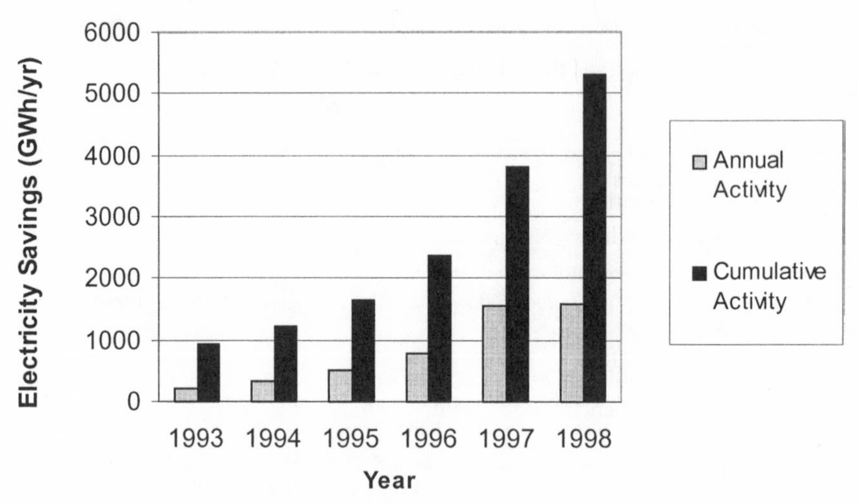

SOURCE: Geller et al. 1999.

FIGURE 4-4: Electricity conservation trends in Brazil.

PROCEL has had other positive impacts besides the direct economic savings. PROCEL contributed to the development of a number of new technologies now manufactured in Brazil, including demand limiters, lighting controls, electronic ballasts for fluorescent lamps, and solar water heaters. PROCEL supported the development of an ESCO industry in Brazil and trained large numbers of energy managers and other professionals. PROCEL has also reduced the risk of power shortages, although not enough to prevent the shortages that occurred in 2001.[2]

PROCEL has cofunded many utility energy efficiency projects. One example was a program in the Jequitinhonha valley, a poor, rural region in the state of Minas Gerais. As of the early 1990s, the region was served by an overloaded transmission line. This led PROCEL and the state utility (CEMIG) to promote greater energy efficiency in the region. The goal, cutting peak load in the region by 9 MW, was achieved by distributing CFLs to households, stimulating the use of peak demand limiters in houses with electric resistance shower water heaters, upgrading the efficiency of street lighting, and encouraging peak demand management by industries and farms. The program cost CEMIG and PROCEL about $3 million but enabled CEMIG to postpone a $25 million transmission line investment while improving energy services to low-income households and small farms.

Electric utility privatization and restructuring is occurring in Brazil in conjunction with formation of federal and state utility regulatory agencies. At the urging of PROCEL and others, the federal regulatory agency (ANEEL) adopted a requirement that distribution utilities spend at least 1 percent of their revenues (representing about $160 million per year) on energy efficiency—both for distribution system loss reduction and for end-use efficiency projects. At least one-quarter of this 1 percent must be spent on end-use efficiency. PROCEL is assisting utilities with preparation of energy efficiency plans, monitoring implementation, and evaluating results on behalf of ANEEL.

This policy was modified by the Brazilian Congress in 2000 with a sizable portion of the 1 percent allocated to research and development (Jannuzzi 2001). But the revised policy still requires distribution utilities to spend at least 0.25 percent of their revenues on end-use energy efficiency programs. Due to the power shortages in 2001, utilities spent about $80 million—approximately 0.5 percent of their revenues—on end-use efficiency programs that year (Villaverde 2001). Promotion and distribution of CFLs was one area where utilities greatly expanded their efforts in 2001, contributing to rapid growth in CFL sales. A lighting industry representative estimates that around 50 million CFLs were sold or distributed in Brazil in 2001 compared to about 14 million the previous year (Roizenblatt 2002).

PROCEL has advocated establishing mandatory efficiency standards on appliances, lighting products, and motors sold in Brazil. After considering this proposal for many years, the Brazilian Congress passed a law in 2001 that directs the executive branch to establish such standards based on analyses of technical and economic feasibility. The law also directs the executive branch to develop mechanisms to increase the energy efficiency of new commercial buildings. The government was implementing this new law with PROCEL's support as of 2002.

To increase its funding base and range of activities, Eletrobras/PROCEL applied for and received a US$43 million energy efficiency loan from the World Bank and complementary US$15 million grant from the Global Environmental Facility (GEF) in 2000. These funds are to be matched by an equal or greater amount of Brazilian funding. The loan supports large-scale implementation of proven energy efficiency measures. The grant supports pilot projects featuring new technologies or delivery mechanisms as well as core activities and capacity building. This is the first World Bank loan solely for the purpose of end-use electricity conservation and demand-side management.

PROCEL suffered some setbacks between 1999 and 2001, however, when leadership of the program changed, staff size was reduced, and the budget was cut. This resulted in slow implementation of the World Bank loan and GEF grant, and other problems. It was particularly unfortunate given the strong growth of PROCEL from 1993 to 1998 and the fact that Brazil faced a severe power shortage in 2001.

In summary, PROCEL demonstrates that a national energy efficiency program can be successful if it has strong government backing and funding, works collaboratively with the private sector and other institutions, and focuses on both technological improvement and market development. The PROCEL experience also shows the value of capacity building and policy advocacy. Finally, PROCEL shows the importance and difficulty of maintaining long-term continuity and growth in a government-based national energy efficiency program.

Netherlands: Industrial Energy Efficiency Improvement

The Dutch government adopted a comprehensive set of policies to stimulate industrial energy efficiency improvements starting in 1990. The primary policy mechanism is formal agreements between the government and specific industrial sectors, known as the long-term agreements (LTA) program. The agreements, which are based on independent assessments of cost-effective savings potential, contain negotiated targets that are legally binding and typically call for a 20 percent average increase in energy efficiency by 2000 relative to efficiency levels in 1989 (Nuijen 1998).[3]

To have a large-scale impact, the government required that at least 80 percent of the firms in a sector participate before an LTA was signed. Participating companies agree to develop an energy efficiency improvement plan and improve energy efficiency wherever technically and economically feasible, to contribute to their sectoral target. Companies also agree to report on progress annually. In return, the government provides detailed energy audits of industrial facilities, tax incentives for investments in energy-efficient technologies and other financial assistance, and protection from mandatory energy efficiency regulations. Independent monitoring is carried out by NOVEM, the Dutch energy agency. Firms that fail to report results or contribute to meeting their sector's LTA can be expelled from the program and subject to tougher energy and environmental standards.

By the end of 1996, the Dutch government signed LTAs with 31 industrial sectors and six service sectors. About 1,000 industrial companies accounting for over 90 percent of industrial energy use in the Netherlands were participating. The average energy efficiency improvement was 12.5 percent by 1996, 17.4 percent by 1998, and 20.4 percent by 1999, meaning the goal of achieving a 20 percent improvement on average by 2000 was realized (Nuijen 1998, van Luyt 2001). A number of important sectors had surpassed their LTA as of 1999 (Table 4-5). So far, no sectors have abandoned their LTA.

An evaluation of the LTA program found that companies devoted more at-

TABLE 4-5

Energy Intensity Reduction Targets and Achievements in Key Industrial Sectors in the Netherlands

Sector	Target for 2000[a]	Achievement in 1995[b]	Achievement in 1999[c]
Chemicals	20	9.3	23
Iron and steel	20	10.8	16
Paper	20	13.2	21
Textiles	20	9.9	21
Cement	20	—	22
Glass	20	12.0	14
Cold storage and refrigeration	25	—	22

NOTES: [a]Target relative to energy intensity in 1989, in percentage reduction.
[b]Achievement during 1989–1995, in percentage reduction.
[c]Achievement during 1989–1999, in percentage reduction.
SOURCES: van Luyt 2001, Rietbergen, Farla, and Blok 1998.

tention to energy management and opportunities for efficiency improvement once they began participating, and that companies were generally satisfied with the program (Nuijen 1998). The effort was estimated to cost the Dutch government $690 million between 1990 and 2000, including tax incentives and other subsidies. At the same time, industries will save this amount on their energy bills *annually* by 2000 (Nuijen 1998). The LTA program also reduced CO_2 emissions by about 6 million metric tons as of 1997, equivalent to about 3 percent of total CO_2 emissions in the Netherlands that year (Gummer and Moreland 2000).

Given the success of the LTAs and the fact that most of the agreements ended in 2000, the Dutch government and industries began adopting new agreements known as Energy Efficiency Benchmarking Covenants in 1999. The Covenants commit companies to adopt "Best Practices" so that they can be among the most energy efficient (top 10 percent) in their sector worldwide by 2012 (van Luyt 2001). In return, the Dutch government and energy service providers offer financial incentives and technical support. Also, the government has pledged not to adopt new energy taxes or CO_2 emissions caps on participating companies (van Luyt 2001). The Benchmarking Covenants are a more individualized approach than the sector-oriented LTA program.

Over two-thirds of energy-intensive industries had signed a Benchmarking Covenant as of 2000 (van Luyt 2001). It is estimated that the benchmarking program will reduce industrial energy use in the Netherlands by 5 to 15 percent by 2012 (Phylipsen et al. 2002). Also, less-energy-intensive companies are being asked to adopt all energy efficiency measures with an internal rate of return of at least 15 percent. A benchmarking committee and verification bureau have been set up to monitor progress.

In summary, the Dutch LTA program demonstrates that a voluntary program can succeed if there is a strong commitment on the part of the government, a supporting framework, and some pressure on the private sector. The Dutch program included both "carrots" and "sticks" that promoted a high level of participation and compliance. The Dutch experience also shows that there continues to be energy savings potential within industries even after the 20 percent reduction in energy intensity achieved during the 1990s.

China: Improved Cookstoves Deployment

Traditional cooking methods with their high levels of fuel use and smoke cause severe environmental, public health, and social problems for vast numbers of households (women and children in particular) in developing countries (Ravindranath and Hall 1995). China has implemented the most sweeping and successful improved cookstoves program in the world. Around 130 million improved stoves were installed in rural areas between 1982 and 1992, meaning that over half of rural households in China obtained an improved stove. Most of these stoves are fueled with biomass, either wood or crop residues. Although there were problems with quality control and durability in the beginning, these problems were overcome. Most of the improved stoves have saved fuel and improved indoor air quality (Smith et al. 1993).

The Chinese National Improved Stove Program used the following policies and strategies to disseminate improved stoves on this massive scale: (1) R&D through a network of research institutions along with independent testing and monitoring of potential stove designs; (2) decentralized training, promotion, and monitoring through Rural Energy Offices, starting in the counties with the greatest need and interest; (3) training and promotion of rural energy companies and technicians that manufacture, install, and service improved stoves; (4) low-interest loans and tax incentives to help these companies financially; and (5) subsidies for stove purchases by low-income households (Smith et al. 1993).

The entire improved stoves program is overseen by the national Bureau of Environmental Protection and Energy (BEPE) and features many actors (Fig. 4-5). The program benefited from the experience and widespread infrastructure established previously to promote biogas digesters and microhydro facilities in China. Stove R&D is conducted at the national, provincial, and county levels

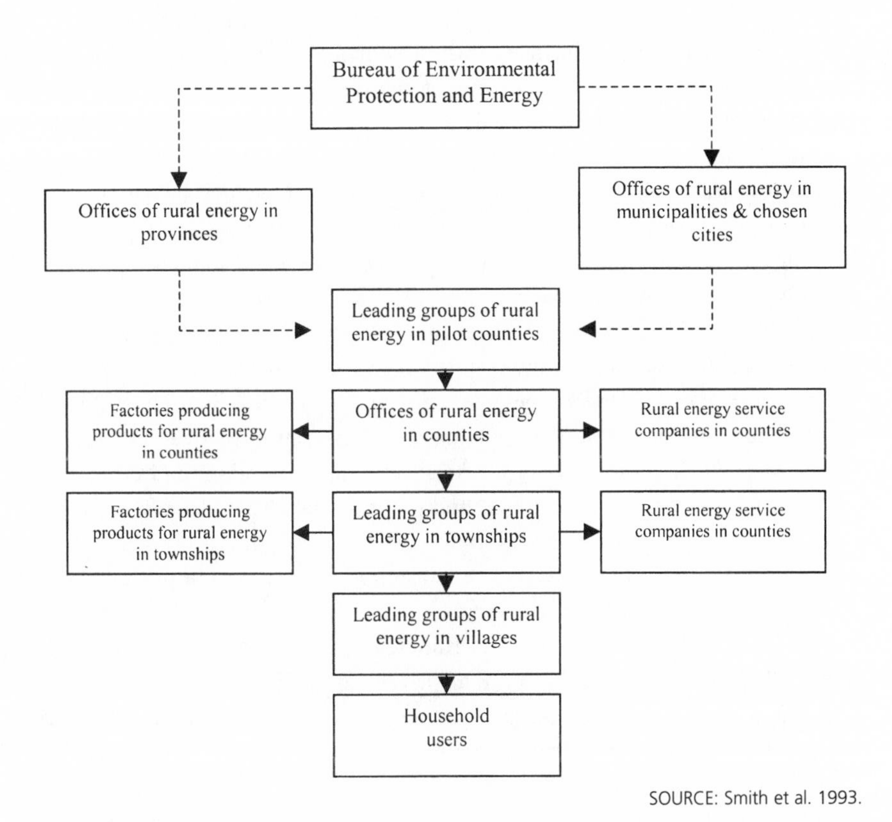

SOURCE: Smith et al. 1993.

FIGURE 4-5: Structure of China's improved cookstove program.

through a network of 25 research institutions funded by the BEPE. Provinces and counties are carefully selected for implementation based on their level of biofuels scarcity, presence of adequate technical, financial and managerial infrastructure, and local interest. Financing, selection of target areas, training of stove technicians and companies, monitoring, and quality control are provided by approximately 1,500 rural energy offices located throughout China.

Improved stoves are built and sold by stove technicians and rural energy service companies, not by governmental agencies. Stoves are sold without direct government subsidy except for free stove parts provided to very poor households in some regions (Smith et al. 1993). About 200,000 workers were involved in stove construction and program support during the 1980s.

Rural households found that fuel savings, improved air quality, convenience, and other benefits justified the roughly $9 average cost for an improved stove. The fuel savings are equal to about 25 percent on average and most stoves have remained in use (Smith et al. 1993). For comparison, about 24 million improved cookstoves were built and installed in India during the 1980s and '90s, but it is estimated that only about 60 percent of these stoves were in use as of 1997 (Ravindranath and Hall 1995, Shailaja 2000).

The Chinese improved stove program was a success in part because of its technical characteristics and quality control. Critical stove parts such as grates and dampers are produced centrally. Stove dimensions such as the size of the fuel opening and combustion chamber have been optimized and standardized, and a flame baffle is used to improve heat transfer to the cooking pots. Chimneys are designed to effectively remove smoke from the kitchen. In addition, stove designs are modified to suit local cooking conditions (Smith et al. 1993).

From 1983 to 1989, the Chinese government (national, provincial, and local) spent about US$158 million supporting the construction of over 110 million stoves. Most of these funds were for local training, promotion, and evaluation; subsidies for low-income households; and program staff. The total cost for building and purchasing these stoves was about $1 billion. Thus the government expenditure was only about 15 percent of the total amount invested in improved stoves.

In summary, China's stoves program provides a number of valuable lessons. First, it showed the importance of coordinated actions at the national, regional, and local levels. Second, it showed that private companies can successfully disseminate energy efficiency or renewable energy technologies to rural households as long as these companies are given training, financial assistance, and marketing support. Third, rural households are willing to pay for an improved energy technology if the benefits are significant and costs are reasonable. And fourth, it showed that technical rigor and quality control can be achieved even in a highly decentralized endeavor such as dissemination of improved cookstoves in rural areas.

United States: More Efficient Electricity Use in California

California has been a leading state in the adoption of appliance efficiency standards, building energy codes, and utility energy efficiency programs. Utilities in the state of California have implemented large-scale energy efficiency programs for over 20 years. As of 2000, investor-owned utilities in California spent about $230 million or 1.4 percent of their revenues on these efforts (Kushler and Witte 2001). The programs include rebates for consumers and businesses that install efficient devices of all types, free retrofits for low-income households, support for implementation of building energy codes, and technical assistance to businesses and industries. Cumulative utility efforts reduced peak demand in the state by 4,400 MW (10 percent) as of 1999 and provided $2.7 billion in life-cycle benefits to California's consumers (Cowart 2001).

The California Public Utilities Commission (CPUC) motivated the utilities to conduct vigorous energy efficiency programs through its regulatory policies. The CPUC allows utilities to recover lost sales revenues and keep a portion of the net societal benefits resulting from their energy efficiency and load management programs. The cost recovery and bonus payments are implemented through a small rate adjustment each year following thorough monitoring and evaluation of each utility's programs. The regulations provide utilities with financial incentives to maximize energy savings and the net economic benefits of their programs (Stoft, Eto, and Kito 1995).

In addition to its utility programs, California has maintained a highly competent state energy agency, the California Energy Commission (CEC), that has implemented a wide range of energy efficiency policies and programs. With authorization provided by state legislation, the CEC adopted appliance efficiency standards and building energy codes starting in the mid-1970s. The CEC has kept these requirements up-to-date through various revisions. Building codes and equipment standards cut electricity use in California by about 12 TWh (5 percent) and saved consumers $1.4 billion per year as of 1998 (CEC 1999). In addition, California's codes and standards paved the way for national appliance efficiency standards and strong building energy codes in other states.

The combination of utility programs, appliance efficiency standards, and building energy codes has had a significant impact on overall energy use in California during the past 25 years. California cut its electricity use per unit of economic output by nearly 30 percent from 1977 to 1999, compared to relatively constant electricity intensity in the other 49 states (Fig. 4-6). California reduced its total energy use per capita by 13 percent and its energy intensity (total energy use per unit of gross state product) by 47 percent from 1970 to 1997 (Geller and Kubo 2000). As of 1997, California's energy use per capita was 32 percent below the national average.

These impressive achievements provided both environmental and economic benefits. One study estimates that by reducing its energy intensity, California lowered its pollutant emissions from stationary sources by over 35 percent as of 1995. The same study indicates that by saving energy and shifting expenditures to more productive areas, California increased its economic output in 1995 by $875 to $1,360 per capita, or 3 to 4.6 percent (Bernstein et al. 2000). However, utilities in California cut their funding for energy efficiency programs in the mid-1990s due to the uncertainty that developed around utility restructuring (CEC 1999).

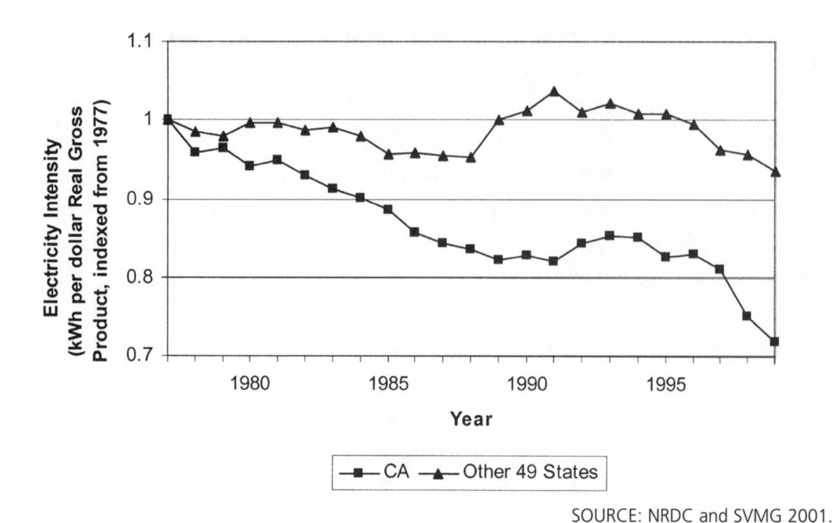

SOURCE: NRDC and SVMG 2001.

FIGURE 4-6: Comparison of electricity intensity trends in California and the rest of the United States.

With the passage of electricity restructuring legislation in California in 1996, policymakers developed a new model for ratepayer-funded energy efficiency programs—a model more appropriate for a competitive electricity marketplace. The legislature mandated a nonbypassable charge of 0.3 cents per kWh (equivalent to about 3 percent of the average tariff) on all electricity sales to fund a variety of "public benefits" activities including energy efficiency programs, low-income energy assistance, renewable energy development, and energy R&D. About half the funds were allocated to energy efficiency programs. Also, the legislation directed utilities to give greater priority to creating a robust energy efficiency services industry in California and to market transformation (Eto, Goldman, and Nadel 1998).

California enacted new legislation in 2000 that extends the public benefits charge through 2012. This will result in about $5 billion of investment in energy efficiency programs, renewable energy measures, and other public benefit activities in this state over the next decade. A 10-year commitment on this scale will provide continuity and will help California maintain its leadership in sustainable energy efforts in the United States.

California updated its building efficiency standards and approved over $500 million for additional energy efficiency programs in 2001. These emergency actions were taken to help the state respond to temporary power shortages and severe electricity price spikes caused by flaws in its utility restructuring policy (Cavanagh 2001, McCullough 2001). Utilities and the CEC increased their financial incentives for energy savings and conducted an extensive energy conservation advertising campaign. In addition, the state began the process of adopting new minimum efficiency standards on refrigeration, lighting, and other products not yet subject to national efficiency standards.

These efforts were very successful. For example, about 33 percent of residential consumers and 25 percent of businesses participated in a program whereby the utilities offered a 20 percent reduction in the electricity rate if the consumer cut electricity use by 20 percent or more. Sales of compact fluorescent lamps (CFLs) increased by over 400 percent between 2000 and 2001 (Calwell et al. 2002). In total, California reduced its electricity use by about 7 percent and peak demand by 10 percent in the summer of 2001 compared to levels the previous year (CEC 2002). These energy and peak demand savings were the main reason California did not experience power shortages that summer. The energy and peak demand savings also contributed to the steep reductions in wholesale electricity prices in the western region (NRDC and SVMG 2001).

In summary, California strongly supported energy efficiency and viewed it as a key "energy resource" for 25 years. The state set up a highly capable energy agency that has adopted leading-edge codes and standards. California has also had well-funded utility energy efficiency programs during most of this period and has given its utilities a financial incentive to operate effective end-use efficiency efforts. As a result, per capita energy and electricity use declined much more sharply in California than in the rest of the United States. This experience demonstrates the importance of consistently supporting energy efficiency through complementary and evolving policies and programs. It also shows that energy efficiency improvement can be an effective response to a short-term energy supply–demand imbalance, especially if there is a solid foundation of programs and service providers to quickly build upon.

India: Renewable Energy Implementation

India established a Ministry for Non-Conventional Energy Sources (MNES) in 1992 to foster the development, demonstration, and deployment of renewable energy technologies. The Ministry supports both rural energy technologies such as improved cookstoves and biogas digesters and modern technologies such as wind power and PV systems. But India actively promoted renewable energy sources well before the Ministry was established (MNES 2001).

The Indian Renewable Energy Development Agency (IREDA) was set up by the government of India in 1987 to finance and promote the manufacturing and adoption of renewable energy technologies. IREDA provides low-interest loans with a 5- to 10-year repayment period. Interest rates vary depending on the type of renewable energy technology. IREDA also supports training in technical and business skills, publicity campaigns, resource assessments, case studies and manuals, business development, and export assistance (Lal 1998). In addition, IREDA helps to get other institutions such as the Industrial Development Bank of India and the Power Finance Corporation to finance renewable energy projects (Mishra 2000).

The Indian government offers one-year depreciation, a 50 percent subsidy on solar home systems, and elimination of import duties and other taxes to further improve the economics of renewable energy adoption (Pachauri and Sharma 1999). The federal government has also encouraged state utilities to provide favorable rates and contract terms for renewable power supplied to the grid by independent project developers. Market development is a major focus of this strategy, which received cofunding from the World Bank and the Global Environmental Facililty in the early 1990s.

This comprehensive market-oriented approach has achieved impressive results (Table 4-6). India is one of the largest users of solar photovoltaic (PV) systems in the world. Also, India ranks fifth in the world with nearly 1,500 MW of installed wind power capacity as of 2001. The total capacity of renewable-based power systems (excluding large-scale hydro projects) was about 3,400 MW at the end of 2001—more than 3 percent of all power generation capacity in the country (TERI 2002). IREDA committed 44 billion Rupees (about US$1 billion at the current exchange rate) for 1,400 renewable energy projects from 1987 to 2000, including financing for over half the installed renewable power capacity (Bakthavatsalam 2001).

TABLE 4-6

Promoting Renewable Energy in India

Policies	Results as of December 2001
Revolving load funds and low-interest loans	1,500 MW of wind power
Accelerated depreciation (tax incentives)	360 MW of bioenergy-based power plants
R&D and demonstrations	1,400 MW of small and microhydropower
Development of manufacturing, marketing, and service infrastructure	50 MW of photovoltaic power systems
Training	3 million biogas plants
Information dissemination	430,000 solar cookers 550,000 square meters of solar water heaters

SOURCES: TERI 2002, Timilsina, Lefevre, and Uddin 2001.

A strong local base for manufacturing and distributing renewable energy technologies has been created in India. Around 75 companies are involved in PV cell and panel production. Over 650,000 solar PV systems aggregating to 50 MW were installed as of 2001 (IEA 2001g). IREDA provides low-interest loans with relatively long loan terms to facilitate the adoption of PV systems, in combination with the government's tax incentives. Also, import duties on PV modules and systems were reduced during the 1990s.

In spite of this progress, IREDA's solar PV financing program has been criticized for its focus on commercial and industrial customers in urban areas who are viewed as less of a credit risk and who can take advantage of the tax incentives. The PV program has also been criticized for its high administrative requirements and poor quality control (Cameron, Stierstorfer, and Chiaramonti 1999, Miller and Hope 2000). It remains to be seen if a network of solar PV supply and service centers as well as functioning microfinance schemes can be set up in off-grid rural areas.

In the area of bioenergy, about 360 MW of biomass-based power systems are operating and another 370 MW of capacity is under construction in India (Bakthavatsalam 2001, TERI 2002). Small and medium-size biomass gasifiers have also been commercialized in India, with about 1,700 units generating the equivalent of 35 MW as of 2000 (Jain 2000). Applications have shifted over time away from irrigation pumping (which requires very high subsidy) toward thermal and power generation due to their more favorable economics.

Over 3 million biogas plants have been installed in India, saving about 3 million metric tons of fuelwood annually and improving the quality of life of plant owners and their families. Adoption of biogas plants is supported by a national program that includes a network of 17 regional biogas development and training centers as well as subsidies equal to one-third to one-half of the digester cost. However, a survey of nearly 500 plants in the mid-1990s found that 19 percent were not in use and another 30 percent had significant defects (Dutta et al. 1997). Consequently, observers have recommended better training of users in operation and maintenance, as well as expanding the network and skills of biogas technicians.

Dozens of companies are engaged in production and assembly of wind turbines in India, often through licensing or joint ventures (Kamalanathan 1998). Most of the installed capacity is privately owned and is located in states where utilities and state energy agencies have provided active support. Installation of wind turbines expanded rapidly between 1994 and 1996, but slowed during the late 1990s (Fig. 4-7). This slowdown is attributed to a reduction in the value of tax credits, reduced support for wind power development in some states, poor site selection in the past in some cases, and poor turbine performance in some

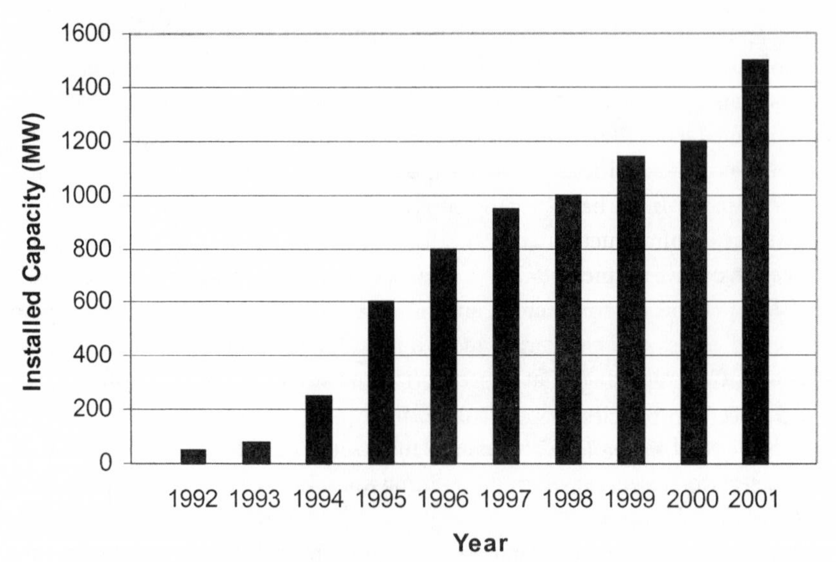

SOURCE: Rajsekhar, van Hulle, and Jansen 1999, Martinot 2001.

FIGURE 4-7: Installed wind power capacity in India.

instances. Capacity utilization averaged only 12 to 15 percent from 1994 to 1996, compared to 30 percent or greater in other countries. Furthermore, some early projects were motivated mainly by very generous tax incentives (Mishra 2000, Rajsekhar, van Hulle, and Jansen 1999).

Due to these problems, renewable energy supporters called for better wind resource assessment, reorienting incentives to reward performance (e.g., paying incentives per kWh produced rather than per rupee invested), improving the design of wind turbines and wind farms (e.g., adapting turbines to Indian power and site conditions), and establishing wind energy technology centers to provide certification, testing, training, and technical support to wind power developers (Rajsekhar, van Hulle, and Jansen 1999). Consequently, the MNES established a wind technology center in Tamil Nadu, started a wind turbine certification program, and prepared site planning and selection guidelines (IEA 2001g). Also, turbine manufacturers have made some technical improvements such as including better power conditioning equipment in newer turbines (Mishra 2000). These efforts seem to be working. Wind power implementation accelerated again including development of a 240 MW project in Maharashtra state in 2001 (AWEA 2002).

Based on the success of renewable energy efforts in India to date, the government has set goals of adding 12,000 MW of renewable-based power capacity to the electric grid, deploying 2 million solar home lighting systems and 1 million solar water heating systems, and building 3 million new biogas digesters by 2012 (Bakthavatsalam 2001). If these goals are met, renewables would provide about 10 percent of new electric capacity additions in India during 2002 through 2012. To meet these fairly ambitious goals, IREDA is expanding its financing and market development activities.

In summary, India has made tremendous strides in developing a renewable energy supply infrastructure and renewable energy markets. The Indian strategy includes attractive financing and financial incentives, technology RD&D, renewable energy business development, and extensive promotion. Thus India is focusing on both the supply and demand side of the renewable energy equation. There have been some "growing pains" related to technology performance as well as the design of certain policies, but India appears to be learning from and correcting these flaws. The future looks promising given the infrastructure that has been created and the high-level commitment to renewable energy in India.

Brazil: Ethanol Fuel

Brazil leads the world in reliance on renewable energy, obtaining 57 percent of its total energy supply from renewable sources as of 2000[4] (MME 2000). Hydropower accounts for about 38 percent of energy supply; energy sources made from sugarcane (ethanol fuel and the sugarcane by-product known as bagasse) account for about 9 percent; and firewood, charcoal, and other renewable sources account for 10 percent. The National Alcohol Fuel Program, known as ProAlcohol, was a key element of Brazil's energy policy during the past 25 years.

Production of ethanol fuel from sugarcane began in 1975 as a way to reduce oil imports and provide an additional market for Brazil's sugar producers. Ethanol production was stimulated through a combination of policies including: (1) low-interest loans for the construction of ethanol distilleries, (2) guaranteed purchase of ethanol by the state-owned oil company at a price considered adequate to provide a reasonable profit to ethanol producers, (3) pricing of neat ethanol so it is competitive if not slightly favorable to the gasoline–ethanol blend, and (4) sales tax incentives provided during the 1980s to stimulate the purchase of neat ethanol vehicles. Price regulation ended in the late 1990s with relatively positive results.

Petrobras (Brazil's national petroleum company) initially opposed the ethanol fuel program, but their opposition was overcome by supporters within and outside of the government. Ethanol production grew rapidly between 1975 and 1985 and then leveled off (Fig. 4-8). Production increased again in the

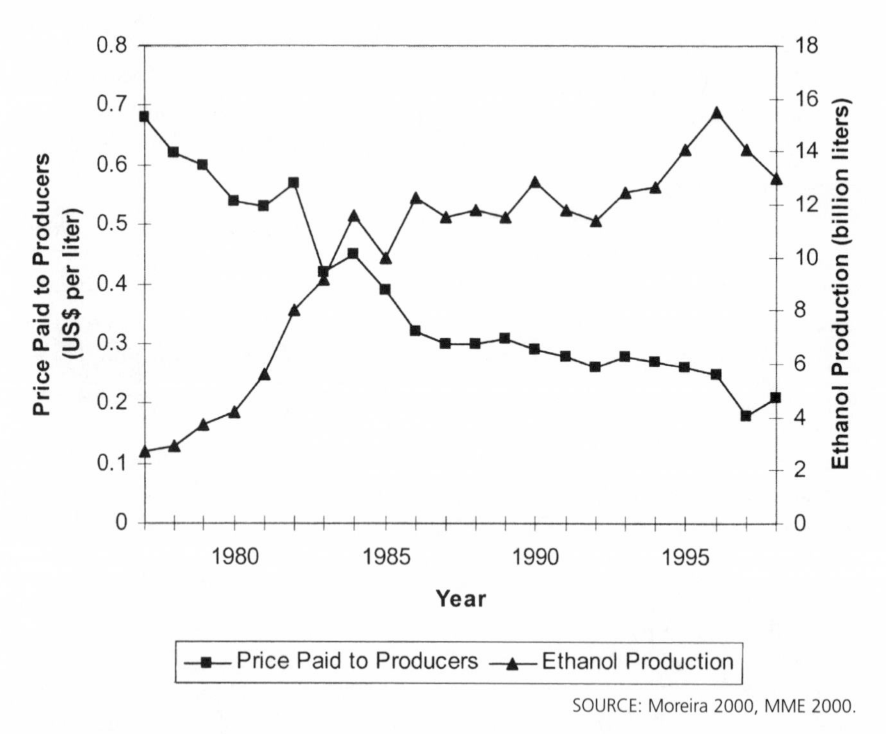

SOURCE: Moreira 2000, MME 2000.

FIGURE 4-8: Ethanol production and price trends in Brazil.

mid-1990s, reaching 13 to 15.5 billion liters per year as of 1997–99 (MME 1999). Sugarcane for ethanol is grown on about 2.7 million hectares, equivalent to about 5 percent of the land area used to grow crops in Brazil (Moreira and Goldemberg 1999). This sugarcane is processed in around 350 privately owned distilleries.

During the first phase of the program (1975–79), the main target was to produce anhydrous ethanol for blending with gasoline using existing distilleries and annexes to existing sugar mills. Ethanol producers received heavily subsidized loans to finance their capital investments (Geller 1985). The goal of achieving 20 percent ethanol in the gasoline blend by 1980 was nearly reached.

The government decided to greatly expand ethanol fuel use in the wake of the second oil price shock in 1979. Research and development at a government-supported laboratory proved that ethanol-fueled cars were feasible. The government and automobile manufacturers signed an agreement that led to large-scale production of neat ethanol cars starting in 1981 (Geller 1985). With strong government support, a large number of autonomous distilleries were built, and large-scale production of hydrated ethanol occurred. Annual production of ethanol fuel quadrupled during the second phase of the program (1979–89).

From 1983 to 1989, the large majority of cars and light trucks sold in Brazil consumed neat ethanol. Demand for these vehicles was created by lowering the sales tax on ethanol-fueled vehicles compared to the tax on vehicles that operate on the gasoline–ethanol blend, and pricing neat ethanol so that it cost drivers slightly less to drive ethanol-fueled vehicles. The government essentially subsidized ethanol fuel through recycling a portion of the substantial tax it collected on gasoline (Moreira and Goldemberg 1999).

The Brazilian government cut the price it paid to ethanol producers throughout the 1980s, particularly after the world oil price collapse in 1986 (see Fig. 4-8). Fuel producers complained and stopped increasing ethanol production by the late 1980s. This in turn resulted in ethanol shortages and the need to import ethanol and methanol starting in 1990. Economic incentives for ethanol vehicles were then cut, leading to relatively low levels of production of cars operating on neat ethanol during most of the 1990s. Alcohol fuel import was no longer needed after 1996.

Following a period of political and economic turmoil in the early 1990s, the Brazilian government emphasized reducing inflation and cutting government expenditures. The government continued to reduce the price paid to ethanol producers, which pressured them to cut costs and improve productivity. The price paid to ethanol producers as of 1996–97 was still about twice the price of gasoline, in spite of nearly a factor of three reduction in the cost of ethanol production between 1980 and 1996. Nonetheless, ethanol production started to increase again in 1996 due mainly to growing fuel demand, continuing cost reductions, and favorable market conditions. Ethanol provided about one-third of the fuel consumed by cars and light trucks in Brazil on an energy basis as of 1998.

In addition to these other policies, ethanol producers in the state of São Paulo established an R&D and technology transfer center. The center has been very effective in improving sugarcane and ethanol yields and reducing production costs. Sugarcane yields and sucrose levels increased as a result of introducing improved sugarcane varieties and better field management. Ethanol yields increased due to better process control, development of continuous fermentation techniques, improved yeast cultures and distillation equipment, and other advances. Also, a benchmarking program in São Paulo state proved to be very important for technology transfer and rapid diffusion of these agricultural and process improvements (Carvalho 1998). The center has helped Brazil become a world leader in sugarcane production and ethanol-conversion technology.

Subsidies for ethanol production and use during the past 25 years were justified based on the positive economic, social, and environmental impacts of the ProAlcohol program. Production of ethanol from sugarcane improves Brazil's balance of payments, reduces unemployment and urbanization pressure, and provides both local and global environmental benefits. Concerning balance of payments, production of ethanol from 1976 to 1996 saved Brazil about $33 billion in oil imports (1996 dollars), or around $50 billion considering that imports would have been partially debt-financed (Moreira and Goldemberg 1999). This is approximately equal to Brazil's hard currency reserves.

Sugarcane and ethanol production support around 700,000 jobs in rural areas, making this sector the largest agroindustrial employer in Brazil (Moreira and Goldemberg 1999). Workers receive reasonable wages, and employment is relatively stable; there is a relatively low index of seasonal work compared to other agricultural sectors in Brazil (Moreira and Goldemberg 1999). Furthermore, the total investment cost per worker in the ethanol industry is much lower than the average for other industries in Brazil, and is on the order of 20 times less than the investment cost per worker in the petrochemical industry (Oliveira et al. 1998).

Concerning environmental impacts, introduction of ethanol fuel reduced lead, sulfur, hydrocarbon, and carbon monoxide emissions from vehicles in Brazil, while nitrogen oxide (NOx) emissions remained about the same. Ethanol fuel has played a significant role in improving urban air quality in the large cities in Brazil (Rosillo-Calle and Cortez 1998). Also, Brazil avoided about 13 million metric tons of carbon emissions due to ethanol substitution for gasoline as of 1996–97, equivalent to nearly 30 percent of its actual carbon emissions from burning fossil fuels (Carvalho 1997). However, burning sugarcane leaves and tops in the field produces local air pollution. Consequently, the government has adopted legislation to gradually phase down preharvest burning (Moreira 2000).

In the late 1990s, the federal government deregulated the price of ethanol, allowing it to be determined by the market. This led to a further decline in the retail price of ethanol, which reached about 30 cents (US$) per liter (36 cents per liter of gasoline equivalent) as of mid-2002. This in turn led manufacturers and consumers to renew their interest in neat ethanol vehicles. Ethanol vehicle production started to expand in 1999, motivated in part by increases in the world oil price in late 1999 and 2000. About 20,000 neat ethanol vehicles were sold in 2001. Also, vehicle manufactures are considering producing dual fuel models that can run on either ethanol or the gasoline–ethanol fuel blend.

Recognizing the economic, social, and environmental contributions of the ethanol fuel program, the Brazilian government adopted some new initiatives to expand ethanol demand starting in 1998. These initiatives included increasing the fraction of ethanol in the gasoline blend to 24 percent as of early 2002, requiring federal agencies to purchase new vehicles that operate on neat ethanol, and testing of the addition of ethanol to diesel fuel. Blends of 3 to 11 percent ethanol in diesel fuel proved to be an effective strategy for reducing particulate emissions from diesel engines (Moreira 2000). Three percent ethanol can be used directly, whereas higher levels of ethanol require another additive, which adds to the fuel cost.

As of 2000, considerable attention was being devoted to finding new uses for sugarcane and ethanol residues, including cogeneration of electricity from bagasse and leaves, production of animal feed, conversion of residues to additional ethanol using acid hydrolysis or enzymatic conversion, and utilization of other sugarcane by-products (Moreira 2000). These initiatives could improve the economic attractiveness of the ethanol fuel program and further expand bioenergy use in Brazil.

Ethanol distilleries already cogenerate steam and electricity for internal use, but they do so at low efficiency because they are unable to sell excess power to utilities or other end users. Regulations are changing to give distilleries (and other businesses) an incentive to produce and sell excess electricity, in part to help avoid power shortages in the future. A new law adopted in 2002 requires distribution utilities to pay 80 percent of the average retail electricity price for excess power from biomass cogeneration projects (as well as wind and small-scale hydro projects) over a 15-year period. This law is expected to result in considerable investment in more efficient bagasse cogeneration systems by ethanol distillery owners (Moreira 2002). Full adoption of higher pressure boilers, more efficient steam turbines, and year-round operation could result in about 31 TWh of biomass-based electricity generation by ethanol distilleries, nearly eight times more electricity than they generated as of 2000 (Moreira, Goldemberg, and Coelho 2002).

The ethanol fuel program will continue to face challenges, particularly as the fleet of neat ethanol vehicles produced during the 1980s is retired. The ethanol fuel mix as of 1999 was 56 percent hydrated ethanol that is used in neat ethanol vehicles and 44 percent anhydrous ethanol that is blended with gasoline (MME 1999). The demand for hydrated ethanol will fall significantly over the next decade unless policies to promote purchase of new ethanol vehicles are strengthened and new outlets such as blending ethanol with diesel fuel are pursued.

In summary, the Brazilian ethanol fuel program is successful for a variety of reasons. First, it began with a strong industry base (namely the existing sugar industry in Brazil) and it worked through the private sector. Second, it has had strong and relatively steady government support, perhaps linked to the large number of jobs created by the program. Third, it was "home grown" and not dependent on foreign technology or financial assistance. Fourth, it featured both substantial financial incentives and market reserves—policies that clearly made a difference. Last but not least, it included significant R&D and ongoing technological improvements.

Denmark: Wind Power Deployment

Denmark initiated a wind energy development program as part of its overall energy plan published in 1976. The initial goal was to cut oil and other energy imports, but protecting the environment also became important in the 1980s (Moore and Ihle 1999). Starting in the late 1970s, the Danish government began RD&D programs and introduced capital subsidies to stimulate wind power development. About $75 million was spent on RD&D and testing from 1976 to 1996, about 10 percent of the Danish energy budget (EIA 1999). This support helped Danish companies such as Vestas and NEG Micon develop better-performing and less costly wind turbines. The Danish government also funded wind resource mapping and a wind turbine certification program that proved to be very helpful as wind power adoption expanded (Krohn 2002a).

The Danish government enacted a subsidy of 30 percent of the investment cost in wind turbines starting in 1979. After steady reductions in the cost of wind turbines and growth of the wind power industry, this subsidy was repealed in 1989 (Moore and Ihle 1999). Financial support shifted to guaranteed payments for wind power production (i.e., an electricity feed-in law). Danish electric utilities were required to purchase output from wind turbines for 85 percent of the retail price of electricity, or about 9 cents per kWh. Also, taxes were placed on fossil fuels, with part of the tax used to subsidize renewable energy sources on account of their superior environmental characteristics. Wind power producers receive about 3.8 cents per kWh from the subsidy pool (Moore and Ihle 1999).

The cost of wind power steadily declined as the technology improved, and production rapidly expanded during the 1990s (Turkenburg 2000). This factor, along with the policies listed above, made wind power cost-effective for both private owners and utilities in Denmark (IEA 1997b). Installed wind power capacity grew from about 300 MW in 1990 to over 2,400 MW as of 2001 (Fig. 4-9). Denmark met its goal of supplying 10 percent of electricity generation from wind power well in advance of the original 2005 target date. In fact, Denmark obtained about 15 percent of its electricity in 2000–01 from wind power (BTM Consult 2001). Emissions of CO_2 in Denmark were reduced by 3.5 million metric tons, SO_2 by 6,500 tons, and NO_x by 6,000 tons in 2001 due to wind power generation (DWIA 2002).

Danish wind turbine companies continue to develop novel designs including

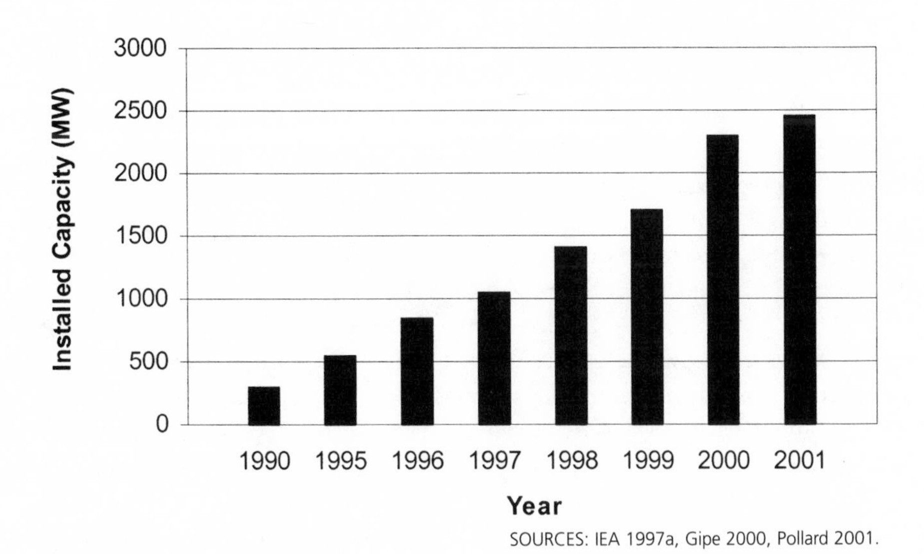

FIGURE 4-9: Installed wind power capacity in Denmark.

very large (2–5 MW) units for offshore applications. The first large offshore wind farm, consisting of 20 turbines (2 MW each), was commissioned in 2001. These innovative turbines have a 72 meter rotor diameter and are expected to produce about 4 GWh/yr under normal wind conditions in Denmark. Utilities were installing two other offshore wind farms of 160 MW each in the North Sea and Baltic Sea as of 2002–2003 (Krohn 2002b).

Denmark's wind power industry is contributing significantly to the Danish economy. About 50 percent of new wind turbines installed worldwide in 2000–2001 were made in Denmark, with the Danish wind turbine industry generating revenues of about $2.7 billion as of 2001. About 20,000 people were employed by the Danish wind power industry as of 2001 (DWIA 2002). Danish wind turbine companies have large market shares in the United States, Germany, and other European countries as well as successful joint ventures in India and Spain. In addition, Danish companies opened a wind turbine assembly plant and a turbine blade manufacturing facility in the United States in the late 1990s—the first new wind turbine facilities built in the United States in many years.

The Danish government has encouraged wind turbine ownership by guilds or cooperatives. About 100,000 Danish families own wind turbines or shares in wind cooperatives, leading to wide public support for wind power and significant political clout through the Danish Wind Turbine Owners Association (Moore and Ihle 1999). The association played an important role in establishing insurance for turbine owners, for example. About 80 percent of all Danish wind turbines are privately owned by cooperatives or farmers.

In recent years the Danish government considered lowering incentive and guaranteed payment levels, and shifting to a Renewable Portfolio Standard (RPS)-type market obligation policy for electricity distributors. But this option was deemed impractical and scrapped indefinitely in 2001. A new conservative government indicated in early 2002 that it will scale back support for renewable energy, but it is unclear if this threat will be acted upon given the strong industrial base, widespread use, and broad support for renewable energy in Denmark (Krohn 2002b). Nevertheless, the uncertainty regarding future payment levels resulted in relatively modest capacity expansion in 2001 (see Fig. 4-9).

In summary, the Danish government nurtured the development of world-class wind turbine designs, a strong manufacturing base, and robust renewable energy markets through a combination of "technology push" and "demand pull." The Danish wind power program demonstrates that ambitious renewable energy goals can be met if there are adequate and consistent financial incentives along with technological and market development. Financial incentives were modified over time as technologies matured, but this was done without compromising the orderly expansion of the wind power market. Also, strong political support for the program was generated by building an industry base and involving a large number of citizens in project ownership.

United Kingdom: Shift from Coal to Natural Gas–Based Electricity

The restructuring that occurred in the electric sector in the United Kingdom during the 1990s produced substantial environmental benefits including a significant reduction in sectoral CO_2 emissions. Starting in 1990, the U.K. power industry was largely privatized and restructured. Competition steadily increased, subsidies for coal production were cut, natural gas production was privatized and deregulated, and regulations to cut emissions of acid rain precursors were implemented (Eikeland 1998).

These policies led to rapid growth in natural gas–fired, combined-cycle generation by privatized utilities as well as independent power producers (the so-called dash to gas). Natural gas achieved a 34 percent market share for power production as of 1999 compared to virtually no natural gas use for power production prior to 1990 (Eyre 1999, Scullion 2001). The fraction of power production from coal fell from 65 percent in 1990 to around 32 percent by 1999 (Fig. 4-10).

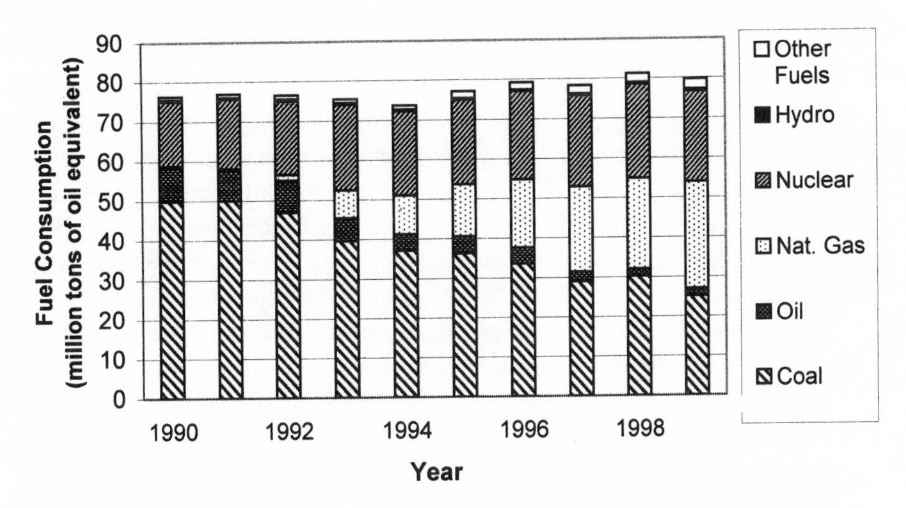

SOURCES: Eyre 1999, Scullion 2001.

FIGURE 4-10: Fuel input for power production in the United Kingdom.

The shift from coal to natural gas for power production reduced carbon emissions by about 14 million metric tons between 1990 and 1997, equivalent to an 8 percent reduction in total U.K. carbon emissions (Eyre 1999). This significant reduction was due to both the lower carbon content of natural gas compared to coal and the efficiency improvement provided by state-of-the-art combined-cycle power generation. Other policies that contributed to a "decarbonization" of power generation in the United Kingdom include improvements in nuclear power plant performance and promotion of combined heat and power (CHP) systems. Also, the Non–Fossil Fuel Obligation (described in Chapter 3) led to about 2.5 percent of capacity supplied by renewables and waste materials (Eyre 1999).

Following the election of a new government in 1997, additional steps were taken to reduce CO_2 and other pollutant emissions. The government set targets of obtaining over 10 percent of total electricity from renewable sources and doubling installed CHP capacity by 2010. A Climate Change Levy on fossil fuels and nonrenewable electricity purchased by businesses was adopted and implemented in 2001. However, a moratorium on approval of new gas-fired, combined-cycle power plants was also adopted due to concerns that further deep and rapid reductions in coal use could occur (Eyre 1999). Ironically, this policy gave a boost to CHP since gas-fired CHP plants are exempt from the moratorium. The moratorium was lifted and there was further replacement of coal-fired capacity with new combined-cycle gas plants in 1999. Natural gas accounted for about 39 percent of electricity generation in the United Kingdom as of 2000 (IEA 2001e).

In summary, the experience in the United Kingdom shows that restructuring and increased competition in the power and fuels sectors can be compatible with environmental protection and declining CO_2 emissions. However, this positive result is by no means guaranteed, as evidenced by the experience in the United States where electric sector restructuring has resulted in greater reliance on low-cost, coal-fired generation.[5] The U.K. government took specific actions, including cutting subsidies for coal mining, adopting stronger emissions standards, and promoting combined heat and power systems, to achieve these positive environmental results in conjunction with energy sector restructuring.

BOX 4-1

Corporate Transformation

Some multinational corporations are starting to make the transition to high levels of energy efficiency and greater reliance on renewable energy sources, driven primarily by the threat of global climate change and policy responses to it. These forward-looking companies see the regulation of carbon dioxide and other greenhouse gases as inevitable and are positioning themselves as leaders as these regulations unfold. These companies also recognize that clean and efficient energy technologies represent a huge market and potential source of future revenues and profits. Here are examples of what some companies are doing.

British Petroleum (BP) pledged to cut its internal greenhouse gas emissions by 10 percent from 1990 levels by 2010. In March 2002, BP announced that it had met this goal eight years ahead of schedule. BP has also set up a renewable energy business division in which it is investing $100 million per year. This business is expected to grow 40 percent in 2002.

Royal Dutch/Shell pledged to reduce its internal greenhouse gas emissions by 10 percent from 1990 levels by 2002. In fact the company reached this target in 2000 through a combination of efficiency improvements, reduced gas flaring, and an internal trading system. Shell has also created two new businesses—Shell Renewables and Shell Hydrogen.

DuPont pledged to reduce its greenhouse gas emissions 65 percent from 1990 levels by 2010. The company has already achieved a 50 percent cut, mainly by changing the way it manufactures nylon (and thereby reducing trace greenhouse gas emissions). DuPont has also pledged to obtain 10 percent of its energy from renewable sources by 2010.

Alcoa pledged to reduce its greenhouse gas emissions 25 percent from 1990 levels by 2010, and more if the inert anode technology it is developing proves to be technically and commercially viable.

IBM pledged to reduce its energy use and associated carbon dioxide emissions by 4 percent per year over the period from 1998 to 2004. IBM has also pledged to increase the energy efficiency of its products and reduce the trace greenhouse gas emissions from semiconductor manufacturing. IBM's worldwide energy efficiency efforts cut its energy use and carbon emissions by nearly 5 percent in 2000.

United Technologies Corp. pledged to reduce its energy consumption as a percentage of sales by 25 percent from 1997 levels by 2007. United Technologies had already cut its energy use per dollar of revenue by 22 percent and its absolute energy use by about 15 percent between 1997 and 2000.

These companies have discovered that large reductions in energy use and greenhouse gas emissions are possible if a commitment is made, systems for identifying and implementing improvements are established, and a process of continuous improvement is put in place. The companies have also discovered that a clean energy strategy can benefit both the environment and their "bottom line."

SOURCES: Dunn and Flavin 2002, Lazaroff 2002, Margolick and Russell 2001.

◉ ◉ ◉

Summary

These case studies show that substantial energy efficiency improvement and renewable energy implementation can be achieved at the national or sectoral scale by adopting a complementary set of policies. The rapid rates of efficiency improvement and renewable energy adoption that occurred in these examples are much greater than the rates experienced in other countries where policies are weak or absent. High levels of efficiency improvement and renewable energy implementation have been achieved in both industrialized and developing countries, demonstrating that any nation can overcome the barriers inhibiting clean energy development through well-designed and well-implemented policy initiatives.

The case studies also show that various combinations of policies can be effective. Some examples included strong regulations or market obligations, others relied on voluntary agreements, financing, pricing reform, or financial incentives as their primary policy instruments. Financial incentives played a key role in many but not all of the case studies. However, all of the case studies involved strong government commitment and support, as well as active participation and implementation by the private sector. In addition, capacity building was a key element of the overall strategy in many of the examples.

The case studies illustrate the importance of sustained efforts for at least 10 years and in some cases more than 20 years. The policies adopted in these countries have evolved over time, as have the technologies themselves. For example, government-funded R&D has been scaled back or financial incentives reduced as markets for efficiency measures or renewable energy technologies evolved and costs declined. In some cases it has been necessary to update regulations or voluntary agreements, or adopt additional policies to maintain efficiency improvements or renewable energy market growth, as markets matured.

The case studies included information on the economic, environmental, and in some cases social benefits of large-scale energy efficiency improvement and renewable energy implementation. Based on these examples, clean energy development provides broad benefits including lower energy bills, increased employment, reductions in pollutant emissions, and reductions in energy imports. The next two chapters present and analyze comprehensive clean energy strategies for two countries—the United States and Brazil. These more detailed analyses of a broad set of policies for advancing energy efficiency and renewable energy at the national level confirm that clean energy development is a "win–win" proposition.

Notes

1. This estimate is based on official gross domestic product (GDP) figures. But GDP growth appears to have been overestimated by the Chinese government. Revised GDP estimates indicate that GDP growth from 1980 to 1999 was closer to 7.6 percent/yr on average than the official figure of 9.8 percent/yr (Sinton and Fridley 2000). Even with the revised GDP growth estimates, energy intensity still fell dramatically in China during the past two decades.
2. The power shortage in 2001 was due primarily to inadequate investment in electricity supply in recent years, as well as to drought and consequently below average hydroelectric capacity in some portions of Brazil.
3. Energy efficiency is based on primary energy use per unit of output; 23 sectors adopted a 20 percent efficiency improvement target, five sectors adopted

a target of less than 20 percent, and nine adopted a target greater than 20 percent.

4. This figure assumes hydropower is valued at the thermal equivalent of fuel oil used to produce the same amount of power in thermoelectric power plants, the convention in Brazil.
5. From 1990 to 2000, 49 percent of the growth in power production in the United States was provided by coal-fired capacity, 28 percent by natural gas, and only 3 percent by non-hydro renewable sources (EIA 2001a). Total carbon dioxide emissions from electricity generation increased by 26 percent in the United States from 1990 to 2000 (EIA 2001c).

The United States: Policies and Scenarios

The United States, with 4.6 percent of the world's population, uses 26 percent of all energy consumed worldwide and 30 percent of all electricity (BP 2001, EIA 2001b). On a per capita basis, the United States consumes 2.5 times more energy than Western Europe, 8 times more than Latin America, 10 times more than China, and 14 times more than other developing countries in Asia and Africa (UNDP 2000).[1] Given these ratios, the United States contributes a disproportionate share of the global problems caused by conventional energy use.

Both affluence and inefficiency contribute to the "hyperconsumption" of the United States. Consider its vehicles and driving habits: Americans have 1.1 passenger vehicles per licensed driver[2]; one in six households have three or more vehicles; an increasing number of trips are made in single occupancy vehicles; the average vehicle is driven 11,800 miles (19,000 km) per year; and collectively Americans drive about 2.5 trillion miles (4 trillion km) per year! To make matters worse, inefficient light trucks (SUVs, minivans, and pickups) now account for nearly 40 percent of all passenger vehicles on U.S. roads (Davis 2001). And with the growing popularity of gas guzzling light trucks, the average fuel economy of new passenger vehicles *declined* over the past 15 years (EPA 2000).

Given these factors, it comes as no surprise that gasoline use in the United States is rising. Consumption of gasoline and other petroleum products increased 15 percent during the 1990s alone (EIA 2001a). The average American consumes 562 gallons (2,130 liters) of gasoline and diesel fuel per year—2.6 times more than the average western European, 3.1 times more than the average Japanese, 15 times more than the average Russian, and 21 times more than the average person living in a developing country (EIA 2001b).

Total energy consumption in the United States rose 45 percent—from about 68 quadrillion British thermal units (quads) in 1970 to 98.5 quads in 2000 (Fig. 5-1). Carbon dioxide (CO_2) emissions due to energy consumption rose about 36 percent—from 1,149 million metric tons (MMT) of carbon in 1970 to 1,562 MMT in 2000 (EIA 2001c). CO_2 emissions rose over 15 percent from 1990 to 2000 alone. Hence the United States did not come close to limiting its CO_2 and other greenhouse gas emissions in 2000 to their level in 1990—a voluntary target established in the Climate Change Convention approved by President George Bush Sr. and the U.S. Senate in the early 1990s.

On the positive side, overall U.S. energy intensity (primary energy per unit of gross domestic product [GDP]) fell 44 percent, 1.9 percent per year on average from 1970 to 2000 (EIA 2001a). Carbon emissions intensity (carbon emissions per unit of GDP) fell 48 percent, 2.1 percent per year on average, since 1970 (Fig. 5-2).

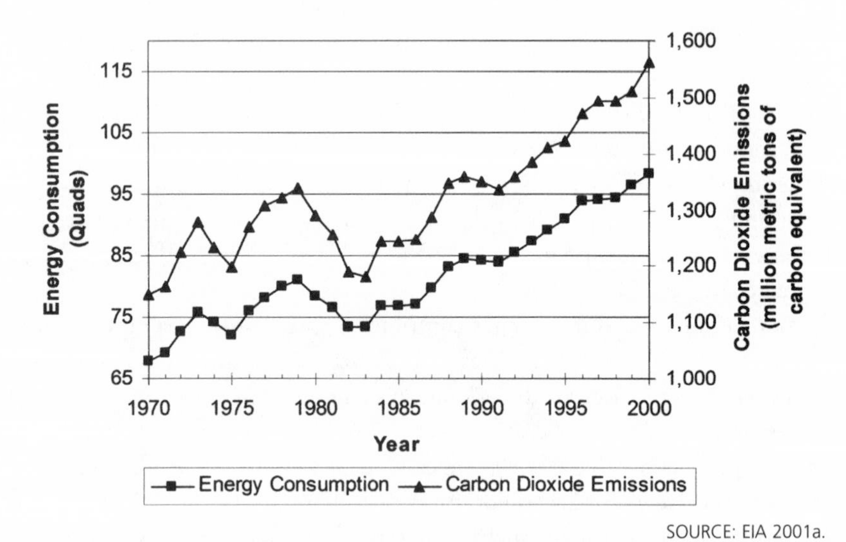

SOURCE: EIA 2001a.

FIGURE 5-1: U.S. energy consumption and carbon dioxide emissions.

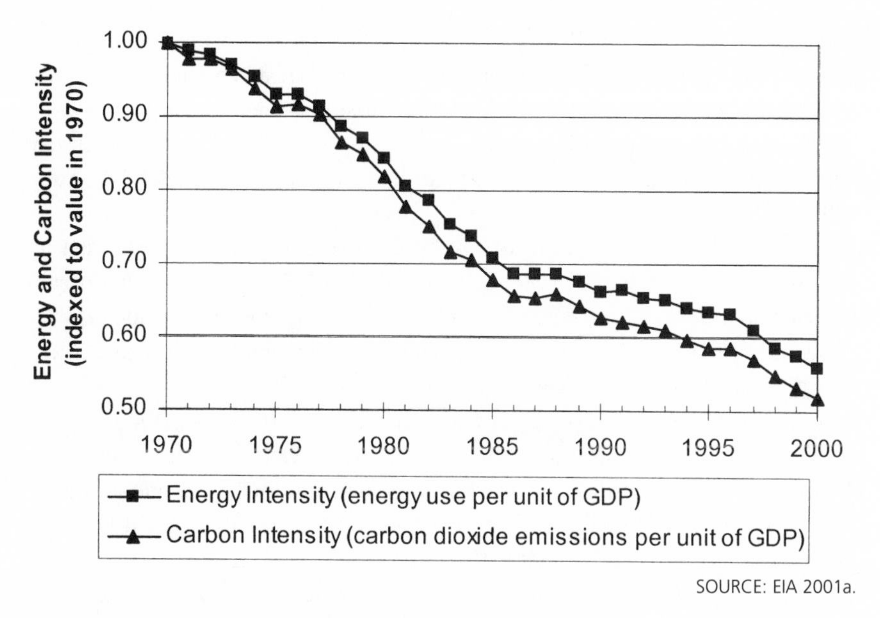

SOURCE: EIA 2001a.

FIGURE 5-2: U.S. energy and carbon intensity.

Carbon intensity declined more than energy intensity because there was a slight shift away from fossil fuels during this period.

If the United States still used as much energy per unit of GDP as in 1970, it would have consumed about 177 quads in 2000 rather than the 98.5 quads actually consumed. This 78.5 quads of savings from efficiency improvements and structural changes is America's most important and underappreciated "energy resource." If only half this growth in energy use had occurred, consumers and businesses would have paid at least $3 trillion more for energy over the past 30 years (Geller and Kubo 2000). Furthermore, it is likely that energy prices would be much higher than they are today if demand growth had not been constrained to the degree it was.

Energy efficiency improvements and structural changes that lower energy intensity also reduce air pollution levels and facilitate meeting emissions standards. Without energy efficiency improvements, there would be greater reliance on older, dirty power plants and industrial facilities. Where total pollutant emissions are capped, efficiency improvements make it easier to comply with emissions caps and reduce the price of emissions allowances.

Reductions in U.S. energy intensity over the past 30 years helped limit growth in greenhouse gas emissions and global warming. U.S. CO_2 emissions as of 2000 would have been about 1,200 MMT greater (in terms of carbon) had energy intensity remained at the 1970 level and fossil fuels provided the large majority of the energy necessary to serve the additional demand. This is equivalent to about a 20 percent increase in *global* CO_2 emissions. The United States would have dumped at least 16 billion more metric tons of carbon into the atmosphere over the past 30 years if energy intensity had not declined (Geller and Kubo 2000).

Three time periods stand out when examining these energy use and intensity trends: (1) 1973 to 1986, (2) 1986 to 1996, and (3) 1996 to 2000. The first period started with the 1973 oil embargo and encompassed the two world oil price shocks. In the mid-1970s, an energy efficiency "movement" was initiated in the energy research and analytical community, as well as in some companies and households. These parties began to view a barrel of oil or kilowatt-hour (kWh) of electricity saved as a resource equivalent to oil or electricity produced. They discovered that there were many ways to save energy at lower cost and lower environmental impact than producing energy (Ford et. al. 1975, Goldstein and Rosenfeld 1976, Ross and Williams 1977, Socolow 1977).

With the growing availability and awareness of energy efficiency measures, U.S. energy intensity fell rapidly—2.6 percent per year on average from 1973 to 1986. About three-quarters of the decline in energy intensity during this period is attributed to efficiency improvements, and about one-quarter to structural change and fuel substitution (Schipper, Howarth, and Geller 1990). Response to oil price hikes, vehicle fuel economy standards, state appliance standards and building codes, and the development and dissemination of new energy-efficient technologies all played a role in the steep decline (Geller et al. 1987, Greene 1999, Vine and Crawley 1991).

The second period, 1986 to 1996, followed the "crash" in the world oil price in 1986. U.S. energy intensity fell just 0.8 percent per year and carbon intensity just 1.1 percent per year during this 10-year period. With these modest rates of reduction, total primary energy use rose 22 percent from about 77 quads in 1986 to 94 quads in 1996 (see Fig. 5-1). Energy intensity continued to fall in some areas, such as space heating, appliances, and passenger vehicles, due to energy efficiency standards, government and utility energy efficiency programs, and the time lags for replacing the stock of vehicles, appliances, etc. But energy intensity leveled off in other areas such as manufacturing and freight transport during the late 1980s and early 1990s (Murtishaw and Schipper 2001).

Beginning in 1997, energy and carbon emissions intensity fell sharply once again. In fact the average rate of energy intensity decline from 1996 to 2000, 3.2 percent per year, was even greater than the rate of decline following the oil price shocks of the 1970s. While the reasons for this recent steep drop in energy intensity are not yet fully understood, it appears that both energy efficiency improvements and structural changes were important. Energy efficiency standards, building codes, and federal, state, and local energy efficiency programs all contributed to the recent drop in energy intensity (Geller and Kubo 2000, Geller, Kubo, and Nadel 2001, EPA 2001), as did the rise of the "information economy" and the shift in manufacturing toward "high tech" industries that are not very energy intensive (Murtishaw and Schipper 2001, Romm, Rosenfeld, and Herrmann 1999). And although oil prices rose sharply in mid-1999, prices for all types of energy were generally falling between 1996 and 2000, making the recent decline in national energy intensity even more remarkable.

It is important to recognize that U.S. energy and carbon intensity did not start falling in 1970. Energy and carbon intensity have declined for centuries as the United States industrialized, moved away from carbon-intensive fuels such as coal, shifted toward a service economy, and advanced technologically. One study estimates that the overall carbon intensity of the U.S. economy declined from about 2.5 kilograms (kg) of carbon per dollar of GDP in 1800 to about 0.36 kg of carbon per dollar of GDP in 1970[3] (Grubler, Nakicenovic, and Victor 1999). This is equivalent to a 1.1 percent average annual rate of reduction. While there has been a long-term trend toward lower energy and carbon intensity, these intensities declined exceptionally rapidly during the past 30 years.

It remains to be seen if recent progress in reducing energy and carbon emissions intensity in the United States can be maintained if not accelerated. A business-as-usual energy future would suggest not, but a more sustainable energy future is possible if a concerted effort is made to increase energy efficiency and shift from fossil to renewable fuels. A careful look at both paths suggests that policy choices can lead to radically different outcomes.

Business-as-Usual

The Bush administration proposes addressing our future energy needs largely by increasing conventional energy supplies—drilling more oil wells, digging more coal mines, and building more pipelines, refineries, power plants, and transmission lines. The Bush administration also proposes billions of dollars in new subsidies for the fossil fuel and nuclear industries, while rolling back some environmental and safety regulations to facilitate development of conventional energy resources (NEPDG 2001). However, this type of energy strategy will be expensive, time consuming (it takes years to develop new large-scale energy sources), and harmful to our environment (NRDC 2001). It is not surprising that the Bush energy plan is highly controversial with the public and Congress—

witness the very contentious debate over opening the Arctic National Wildlife Refuge (ANWR) to oil development.

Continuing current energy policies and trends, and supplementing them with the types of proposals in the Bush energy plan, results in relatively limited energy efficiency improvements, significantly greater consumption of fossil fuels, and only modest increases in renewable energy production in the United States. This type of energy future leads to higher energy bills for consumers and businesses, greater dependence on oil imports, rising CO_2 emissions, and increased global warming (NRDC 2001).

To illustrate the implications of business-as-usual energy policies and trends, it is useful to examine the official energy forecast produced by the Energy Information Administration (EIA).[4] The Reference Case forecast contained in EIA's *Annual Energy Outlook 2001* projects that total U.S. energy consumption will grow 32 percent, 1.3 percent annually, from 1999 to 2020 (EIA 2000c). Oil consumption would increase by about one-third by 2020 in the Reference Case. And with domestic oil production falling, U.S. oil imports would increase by more than 60 percent by 2020. This means the United States would become much more dependent on oil imports from the Organization of Petroleum Exporting Countries (OPEC) and the Persian Gulf region, harming national security and leaving the U.S. economy more vulnerable to future oil price shocks.

Electricity consumption would increase about 45 percent by 2020 in the Reference Case forecast. Approximately 1,260 new power plants will be needed during the next two decades, assuming the typical new plant is 250 megawatts (MW) in capacity. This is consistent with the Bush energy plan, which calls for building 1,300 new power plants in the next 20 years, more than one per week. It is envisioned that most of the additional electricity would come from natural gas–fired power plants, but there would be a 25 percent increase in electricity generation from coal-fired power plants as well. Only 4 percent of the growth in electricity supply would come from renewable sources in EIA's Reference Case. By 2020, all renewable power sources would provide only about 8 percent of U.S. electricity supply—*less than* the 10 percent they provide today.

As noted previously, continuing current policies and trends implies relatively limited energy efficiency improvements. For example, residential energy use per household is projected to increase slightly in EIA's Reference Case forecast, as is energy use per unit of floor area in the commercial sector. Likewise, the Bush energy plan pays lip service to improving energy efficiency, but would do relatively little to actually increase efficiency. In short, the United States will continue to waste energy on a grand scale under a business-as-usual scenario.

With growing energy consumption and increases in some energy prices, national energy expenditures rise nearly 45 percent between 1999 and 2020 in the

Reference Case. However, this forecast assumes very modest increases in oil and natural gas prices, and falling electricity prices, in spite of high growth in demand. If energy demand were to grow at the rates projected by EIA, energy prices could substantially rise, in which case consumers and businesses would pay much higher energy bills than they do today.

Not surprisingly, U.S. CO_2 emissions will continue to climb if total energy use grows and there is only minimal development of cleaner energy sources. The EIA Reference Case forecast projects that CO_2 emissions will rise by 35 percent between 1999 and 2020. The additional emissions of 530 MMT of carbon by 2020 is greater than the combined emissions of Africa, Central America, and South America today. If this happens, the United States would maintain its highly disproportionate share of global CO_2 emissions, in essence "thumbing its nose" at the other nations that are committed to cutting their emissions in accordance with the Kyoto Protocol. And if emissions by the world's largest emitter continue to climb, it will be very difficult to obtain cooperation from developing countries in the worldwide effort to reduce greenhouse gas emissions and limit global warming.

A More Sustainable Energy Future

Energy waste, growth in oil imports, and steadily rising CO_2 emissions are not preordained. The United States, if it so chooses, could significantly increase energy efficiency and renewable energy use, relative to the levels expected under a business-as-usual energy future. The technologies are proven and readily available. But it will take comprehensive and strong policies to overcome inertia and move the United States off the path of rising energy use, rising oil imports, and rising CO_2 emissions.

A more rational energy strategy would first increase energy efficiency to the maximum extent that is cost-effective and achievable. It would then rely primarily on renewable energy sources to satisfy any remaining new energy demand. Energy efficiency improvements, renewables, and some growth in natural gas consumption would be used to reduce consumption of the most problematic conventional energy sources, namely coal, oil, and nuclear energy.

Even though the United States is much more energy efficient today than it was 30 years ago, there is still enormous potential for additional cost-effective energy savings. Some well-established measures such as high-efficiency lighting and appliances are used in less than one-quarter of feasible applications. Other newer energy efficiency measures such as hybrid gasoline–electric vehicles and sealing home heating ducts have barely begun to be adopted. A group of national laboratories estimates that increasing energy efficiency throughout the economy could cut national energy use by 10 percent or more in 2010 and approximately 20 percent in 2020, with net economic benefits for consumers and businesses (Interlaboratory Working Group 2000). And this study, coming from national laboratories affiliated with the Department of Energy (DOE), contains many conservative assumptions.

Renewable energy sources, such as wind power, solar thermal and electric technologies, bioenergy sources, and geothermal energy, are rapidly improving in performance and declining in cost (Short 2002, Turkenburg 2000). Wind and solar photovoltaic (PV) power are the fastest-growing energy sources in the world. As noted in Chapter 3, some states are aggressively pursuing renewable energy sources through renewable portfolio standards, systems benefit charges, and other policies. With adequate policy support at the national level, renewable energy production could rapidly expand (Clemmer et al. 2001, Interlaboratory Working Group 2000).

Policy analysts often create scenarios to study the impacts of different policy options or combinations of options. The author of the present volume presents a Clean Energy Scenario consisting of the 10 energy policies listed in Table 5-1. The Clean Energy Scenario is analyzed and contrasted with a Base Scenario (similar to the EIA Reference Case Forecast) later in this chapter. In short, the 10 policies would substantially increase U.S. energy efficiency and renewable energy production, thereby providing economic, environmental, and national security benefits to American consumers and businesses.

The policies included in the Clean Energy Scenario address the wide range of barriers described in Chapter 2 and cover the full spectrum of policy options described in Chapter 3. The policy set includes R&D to expand and improve energy efficiency and renewable energy technologies, financial incentives and awareness

TABLE 5-1

Policies Included in the Clean Energy Scenario
1. Increase passenger vehicle fuel economy standards
2. Establish a national system benefits trust fund
3. Adopt voluntary agreements to reduce industrial energy use
4. Establish a renewable portfolio standard for power generators
5. Adopt new appliance efficiency standards and stronger building codes
6. Provide tax incentives for innovative renewable energy and energy-efficient technologies
7. Expand federal R&D and deployment programs
8. Remove barriers to combined heat and power systems
9. Strengthen emissions standards on coal-fired power plants
10. Establish renewable energy or carbon content standards for vehicle fuel

building to stimulate commercialization and adoption of these measures, and standards and market obligations to guarantee a high level of implementation. None of the policies by themselves is adequate, but taken together they have the ability to transform markets in an orderly and mutually reinforcing manner.

New energy taxes or a CO_2 emissions tax are not among the 10 policies in the Clean Energy Scenario. Such taxes are opposed by a majority of the public and are not needed to place the United States on a path to a sustainable energy future. But new energy or emissions taxes could reinforce these other policies, although the tax would need to be relatively large to have a significant direct impact on energy efficiency and fuel choice. For example, adopting a carbon tax of $50 per ton by itself (which would lead to about a 10 percent increase in the retail price of gasoline and electricity on average) reduces national energy use by only about 2.5 percent (Interlaboratory Working Group 2000). However, even a small tax could be influential if some or all of the revenue were "recycled" into energy efficiency and renewable energy programs (Bernow et al. 1997).

The 10 policies are presented next, followed by a discussion of their potential impacts.

Increase Passenger Vehicle Fuel Economy Standards (Policy 1)

The original Corporate Average Fuel Economy (CAFE) vehicle fuel efficiency standards were adopted in 1975 and reached their maximum level in 1985. The standards were very successful in cutting U.S. gasoline consumption with few negative side effects, as discussed in Chapter 4. Higher CAFE standards have been recommended by energy analysts since the early 1980s (Ross and Williams 1981) and should be adopted before the United States faces another oil price shock or supply crisis.

This policy calls for increasing the CAFE standards for cars and light trucks by 5 percent per year for 10 years so that they reach 44 mpg for cars and 33 mpg for light trucks in 2012, with further improvements beyond 2012. Alternatively, the standards for cars and light trucks could be combined into one value for all new passenger vehicles, specifically 38 mpg by 2012. This level of fuel economy improvement is technically feasible and cost-effective for consumers, and can be achieved without compromising vehicle safety (DeCicco, An, and Ross 2001). In fact, driving in general would be safer if tougher fuel economy standards induced manufacturers to reduce the size and weight of larger vehicles (Friedman et al. 2001).

Can Detroit do it? The Ford Motor Co. has voluntarily committed to increasing the fuel economy of its new sport utility vehicles (SUVs) by 5 percent annually between 2001 and 2005. General Motors responded that it will exceed Ford's level of fuel economy for its SUVs and other light trucks (Bradsher 2000). If this rate can be achieved in SUVs, it can be achieved in all new vehicles made by Ford, GM, and other manufacturers.

Higher fuel economy standards should be complemented by (1) tax credits for purchasers of innovative, highly efficient vehicles (see Policy 6 below); (2) higher taxes on gas-guzzling vehicles; (3) expanded labeling and consumer education; and (4) vigorous R&D on fuel-efficient, low-emissions vehicles (see Policy 7 below). This combination of policies would facilitate implementation of the tougher standards.

The CAFE standards proposed here would save about 1.0 million barrels of petroleum per day by 2010 and 3.6 million barrels per day by 2020, equivalent to 2.1 and 7.7 quads of energy on an annual basis in 2010 and 2020, respectively.[5] Over 40 years, increasing vehicle efficiency as suggested here would save 10 to 20 times more oil than the projected supply from the Arctic National Wildlife Refuge (ANWR) and more than three times the total proven oil reserves in the United States (Geller 2001).

Establish a National System Benefits Trust Fund (Policy 2)

Many electric utilities operate programs to encourage more efficient energy use, often as a result of state legislation or regulatory mandate. Experience with utility energy efficiency programs in California, the Pacific Northwest, New York, and New England shows that these programs have been highly effective. The value of energy bill savings for households and businesses is two to three times the total cost to produce these savings, and long-term growth in electricity demand is significantly reduced (Nadel and Kushler 2000).

Unfortunately, increasing competition and restructuring have prompted many utilities to cut their energy efficiency programs over the past five years. Total utility spending on all demand-side management programs (i.e., energy efficiency and peak load reduction efforts) fell by more than 50 percent from a high of $3.1 billion in 1993 to $1.4 billion in 1999 (EIA 2000d, Nadel and Kushler 2000).

To ensure that energy efficiency programs and other public benefits activities continue following restructuring, 20 states have established "system benefits funds" through a small surcharge on all kWh flowing through the transmission and distribution grid (Nadel and Kushler 2000). A national system benefits trust fund would provide matching funds to states for energy efficiency programs, assistance to low-income households, and renewable energy development. Specifically, this policy calls for a surcharge of 0.2 cent/kWh—about 3 percent of the average retail price of electricity today—to support these activities.

This policy would give states and utilities a strong incentive to expand their energy efficiency, renewable energy, and low-income assistance programs. All states and utilities would pay into the fund, but they would only get money back if they establish or continue energy efficiency programs and other public benefit activities. However, individual states, not the federal government, would decide how the money would be spent.

This policy should lead to widespread energy efficiency improvements in lighting, appliances, air conditioning, motors systems, and other electricity end-uses. Savings could reach nearly 300 terawatt-hours (TWh) in 2010, 7 percent of otherwise projected electricity use. By 2020, annual savings could exceed 800 TWh, equivalent to 6.5 quads of primary energy. With these levels of electricity savings, the risk of power shortages in the future would diminish, there would be fewer price spikes caused by periods of tight supply and demand, and there would be much less need to build contentious new power plants.

Adopt Voluntary Agreements to Reduce Industrial Energy Use (Policy 3)

There is substantial potential for cost-effective efficiency improvement in industry. For example, in-depth analyses of specific energy efficiency technologies for the iron and steel industry found a total cost-effective energy savings potential of 18 percent (Worrell, Martin, and Price 1999). A similar analysis for the paper and pulp industry found a cost-effective energy savings potential of 16 to 22 percent. Furthermore, new energy-saving industrial technologies and practices continue to be developed and commercialized (Martin et al. 2000).

To stimulate widespread energy efficiency improvements in the industrial sector, this policy calls on the U.S. government to establish voluntary agreements with individual companies or groups of companies. Companies or sectors would pledge to reduce their overall energy and carbon emissions intensities (energy and carbon per unit of output) by a significant amount, for example by at least 2 percent per year over 10 years. Companies that make and follow through on this commitment could be given Energy Star® or similar recognition. The government would encourage participation and support implementation by (1) providing technical assistance to companies that request it, (2) offering to postpone consideration of mandatory emissions reductions or tax measures if a large percentage of industries participate and achieve their goals, and (3) expanding federal R&D and demonstration programs for sectors with high participation.

A number of major companies have already set voluntary goals for improving energy efficiency. In 1995, Johnson and Johnson set a goal of reducing energy costs 10 percent by 2000 through adoption of "best practices" in its 96 U.S. facilities. As of April 1999, they were 95 percent of the way toward this goal, with the vast majority of projects providing a payback of 3 years or less (Kauffman 1999). British Petroleum set a goal of reducing its carbon emissions 10 percent below 1990 levels by 2010, representing nearly a 40 percent reduction compared to projected emissions under a "business-as-usual" scenario. BP met this goal in 2002, eight years ahead of schedule (Lazaroff 2002). DuPont also established voluntary goals to reduce energy intensity, increase renewable energy use, and cut greenhouse gas emissions. If these three companies can do this, so can other companies.

Voluntary agreements between government and industry along the lines proposed here have resulted in substantial energy intensity reductions in Germany, the Netherlands, and Denmark (the Dutch experience was reviewed in Chapter 4). A key factor in the success of these programs was the threat of new taxes or regulations if a large fraction of industries failed to make and comply with voluntary commitments (Price and Worrell 2000).

The impacts of this policy are based on a detailed analysis of potential voluntary industrial agreements in the United States carried out by a team from national laboratories (Interlaboratory Working Group 2000). Based on this analysis, it is assumed that voluntary agreements and supporting activities could reduce industrial energy use by 8.5 percent by 2010 and 16 percent by 2020, relative to levels otherwise forecast.

Establish a Renewable Portfolio Standard for Power Generators (Policy 4)

Utilities and other power generators can be required to supply or purchase a specified amount of capacity or percentage of total electricity generation from renewable sources. This type of requirement is known as a Renewable Portfolio Standard (RPS). Power generators would be allowed to achieve the goal through installation of renewable technologies on their own and/or purchase of tradable renewable energy credits. The tradable credit scheme is designed to minimize the overall cost of compliance. As noted in Chapter 3, 12 states had adopted some type of RPS for their utilities as of 2001. Eligible renewable energy sources generally include wind, solar, geothermal, and biomass-based power, but not existing large-scale hydropower projects (see Table 3-6).

Some members of Congress have introduced RPS proposals, notably Senator Jeffords (I-VT) whose bill The Renewable Energy and Energy Efficiency Investment Act of 2001 (S. 1333) requires 10 percent nonhydro renewables by 2010 and 20 percent renewables by 2020. The U.S. Senate included a modest national RPS in the energy bill it adopted in April 2002. This RPS requires 4 percent renewables by 2010 and 10 percent by 2020, but excludes all public power companies such as the Tennessee Valley Authority, municipal utilities, and rural electric cooperatives from the requirement. For comparison, nonhydro renewables provided about 2.2 percent of U.S. electricity supply as of 2000 (EIA 2001a). This fraction is projected to remain under 3 percent through 2020 in the EIA Reference Case forecast (EIA 2000c).

The RPS in the Clean Energy Scenario is based on Senator Jefford's RPS proposal and applies to all electricity generators (so-called independent power producers as well as traditional utilities). It is assumed that most of the incremental electricity is provided by wind power, which is expanding rapidly and becoming increasingly competitive with conventional sources of electricity. Biomass-based and geothermal power also provide significant amounts of electricity by 2020 in this scenario.

Adopt New Appliance Efficiency Standards and Stronger Building Codes (Policy 5)

Appliance standards and building codes are clearly among the United States's most effective energy-saving policies. This policy calls for the adoption of new efficiency standards for distribution transformers, exit signs, traffic lights, and torchiere lighting fixtures. California has already adopted standards on all these products; Massachusetts and Minnesota have adopted standards on distribution transformers. This policy also calls for efficiency standards on refrigeration and heating equipment used in the commercial sector, as well as on the standby power consumption for electronic products such as televisions, VCRs, cable boxes, and audio equipment. In addition, this policy assumes that new efficiency standards for residential central air conditioners and heat pumps adopted by the Clinton administration are allowed to take effect (the Bush administration has rolled back these standards, but this decision is being challenged in court).

These actions would save approximately 95 billion kWh of electricity in 2010 and 265 billion kWh in 2020. The latter value amounts to about 8 percent of projected residential and commercial electricity use in 2020 in the absence of other policies to raise efficiency. In addition, it is estimated that these standards would have a benefit–cost ratio of about 5 and would yield net benefits of about $80 billion for consumers and businesses (Kubo, Sachs, and Nadel 2001).

Over 20 states have not yet adopted mandatory energy codes for new residential and commercial buildings, or have out-of-date codes (BCAP 2001). This policy would direct states to review their building energy codes and strengthen them if necessary so that all states have "good practice" building energy codes in place by a certain date. The DOE would continue to provide technical assistance for these efforts, with preference given to states that adopt statewide mandatory codes at or above the level of the model International Energy Conservation Code. Furthermore, the model code should be updated periodically, and all states should revise their energy codes after this occurs. Taking these steps could yield 0.3 quads of energy savings in 2010 and 1.5 quads of savings in 2020 (Nadel and Geller 2001).

Provide Tax Incentives for Innovative Renewable Energy and Energy-Efficient Technologies (Policy 6)

Certain renewable energy technologies, mainly wind power projects, are eligible for a federal production tax credit of 1.7 cents per kWh generated over the first 10 years of the project's life. This policy would extend the existing renewable energy production credit and expand it to cover all renewable power technologies other than hydropower. This would help to level the playing field between renewable energy, fossil fuel, and nuclear power sources, with incentives for renewables justified in part by their superior environmental characteristics (Clemmer et al. 2001). However, no additional renewable energy production is assumed because the RPS (Policy 4 above) already results in rapid expansion of renewable power generation.

Many new energy-efficient technologies have been commercialized in recent years or are nearing commercialization. But these technologies may never get manufactured on a large scale or be widely used due to their initial high cost, lack of consumer awareness, and market uncertainty. Tax incentives would help manufacturers justify mass production and marketing of innovative energy-efficient technologies. Tax credits would also help buyers (or manufacturers) offset the relatively high first-cost premium in the early years of production, thereby contributing to sales and market growth. Once the new technologies are produced on a significant scale, costs should decline and the tax credits can be phased out.

This policy includes tax incentives for approximately five years for advanced, high-efficiency appliances, highly efficient new homes and commercial buildings, hybrid and fuel cell vehicles, and combined heat and power (CHP) systems. The total cost to the Treasury would be on the order of $10 billion. These credits would save energy directly, due to purchases of equipment eligible for the credits.

But if the credits help to establish these innovative products in the marketplace and reduce their first-cost premium, the indirect impacts after the credits end would be many times greater than the direct impacts (Quinlan, Geller, and Nadel 2001). The comprehensive energy bill adopted by the U.S. Senate in April 2002 includes tax credits along these lines.

Expand Federal R&D and Deployment Programs (Policy 7)

As noted in Chapter 3, energy efficiency programs funded by the U.S. DOE and Environmental Protection Agency (EPA) contributed to the development and widespread use of many innovative energy efficiency technologies. DOE-funded R&D also helped to reduce the cost and improve the performance of wind, solar PV, and other renewable energy technologies (EERE 2000). Furthermore, President Clinton's Committee of Advisors on Science and Technology (PCAST) recommended substantially greater funding for R&D on energy efficiency and renewable energy technologies due to their potential benefits and the fact that private companies tend to underinvest in R&D (PCAST 1997).

Based on specific budget recommendations in the PCAST report, this policy would increase funding for DOE's energy efficiency R&D programs by 15 percent per year and renewable energy R&D programs by 20 percent per year for at least three years. Likewise, funding for EPA's energy-related greenhouse gas emissions reduction programs would increase by 20 percent per year. The total increase in funding for federal energy efficiency and renewable energy programs would be about $190 million per year.[6]

It is assumed that expanding energy efficiency R&D and deployment programs along these lines would save about 1 quad in 2010 and 3 quads in 2020 (Nadel and Geller 2001). These savings estimates are conservative, assuming that some savings are counted already under other policies recommended here. Furthermore, no additional renewable energy generation is assumed because of the separate renewable electricity and fuels standards (Policies 4 and 10). Expanding R&D on renewable energy technologies, however, would facilitate implementation of and compliance with these requirements.

Remove Barriers to Combined Heat and Power Systems (Policy 8)

CHP systems produce multiple usable energy forms (e.g., electricity and steam) from a single fuel input. These systems achieve much greater efficiency than the usual separate production of steam and electricity. But several flaws in government and utility regulations hinder CHP development. These include (1) environmental standards that do not recognize the efficiency gains of CHP systems (i.e., CHP systems are not given credit for the emissions they offset from utility power plants), (2) utility rules that make it difficult for CHP systems to connect to the utility grid, (3) utilities charging businesses that install CHP systems unfair "exit fees" or onerous rates for backup power, and (4) tax depreciation rules that penalize many types of CHP systems (Casten and Hall 1998).

This policy would address all of these problems. It would reform environmental regulations for CHP systems to give them credit for the avoided emissions from utility power plants. It would enact uniform standards for interconnecting CHP systems with the utility grid. It would prohibit utilities from charging businesses exit fees if they want to install CHP systems, and it would require utilities to charge fair rates for backup power. Finally, it would set depreciation periods for CHP systems based on their actual life.

The U.S. DOE and EPA have established a goal of adding 50,000 MW of new CHP capacity by 2010, roughly equivalent to the amount of CHP capacity in the United States as of 2000. This goal is achievable if the actions recommended here are taken. Furthermore, it is estimated that an additional 95,000 MW of CHP capacity could be added over the period from 2011 to 2020 in response to these actions (Nadel and Geller 2001). Most of this capacity would be fueled with natural gas and could result in net energy savings of about 1.1 quads in 2010 and 2.9 quads in 2020.

Strengthen Emissions Standards on Coal-fired Power Plants (Policy 9)

Many older coal-fired power plants are "grandfathered" under the Clean Air Act. This means that they are not required to meet the same emissions standards for nitrogen oxides (NO_x), sulfur dioxide (SO_2), and particulates as power plants built after enactment of the Clean Air Act of 1970. Currently, 850 power plants built before 1970 are still operating. These plants together have a power capacity of about 145,000 MW and produced about 21 percent of electricity generation in the United States as of 1999 (Shoengold 2001).

Older, dirtier power plants emit 3 to 5 times more pollution per unit of power generated than newer coal-fired power plants, and 15 to 50 times more NO_x and particulates than new combined-cycle natural gas power plants (Cavanagh 1999). These older coal-fired plants are also less efficient than most new plants. When the Clean Air Act was adopted, it was expected that these dirtier power plants would eventually be retired. However, many utilities are extending the life of these older plants indefinitely due to their low operating cost.

This policy would require older power plants to meet the same emissions standards as new plants. Some plants would be modernized and cleaned up, but many would be shut down and replaced with much more efficient and cleaner generating sources such as combined-cycle natural gas power plants or renewable power technologies. It is assumed that this policy would be enacted soon but that the requirements would be phased in over the next two decades. This lead time would allow utilities to evaluate their options and achieve the standards without unduly disrupting power markets.

Alternatively, the same general objective could be achieved by adopting new emissions standards as part of a Clean Air Act "four pollutant" strategy. This strategy has been proposed to address SO_2, NO_x, mercury, and CO_2 emissions in an integrated fashion, for example in bills introduced by Senators Jeffords and Lieberman (S. 556) and Representatives Boehlart and Waxman (H.R. 1256) in the 107th Congress. This strategy includes tradeable CO_2 emissions allowances, with emissions credits provided for the shutdown of older coal-fired plants. CO_2 emissions credits along with strict emissions caps on all four pollutants would lead to the retirement of some older coal plants.

It is assumed that this policy, along with the RPS, would lead to replacement of 25 percent of the electricity generated by older coal-fired plants with electricity from new natural gas-fueled combined cycle plants or renewable facilities by 2010. By 2020, it is assumed that 50 percent of the generation from older coal plants would be replaced. Due to the higher efficiency of the new gas-fired plants, about 0.9 quads of energy would be saved by 2010 and 1.8 quads by 2020. Furthermore, this type of policy would significantly reduce air pollutant emissions (ELI 2000).

Establish Renewable Energy or Carbon Content Standards for Vehicle Fuel (Policy 10)

Fuels derived from renewable energy, such as ethanol produced from biomass or hydrogen made from solar energy, are under active development. The large scale and successful ethanol fuel program in Brazil was reviewed in Chapter 4. In the United States, about 1.6 billion gallons (6 billion liters) of ethanol fuel were produced from corn as of 2000 (EIA 2001a). This fuel is blended with gasoline, thereby reducing petroleum consumption and emissions of some pollutants.

Gasoline suppliers could be required to achieve a minimum renewable energy content in their fuel supplies, similar in concept to the renewable portfolio standard for electricity generation. The standard could be specified as a minimum renewable energy content or as a cap on the average CO_2 emissions factor of vehicle fuels. Fuel suppliers could have the flexibility of meeting the standard on their own or buying tradable credits from other producers of renewable or low-carbon fuel.

A renewable fuels standard is included in a number of bills introduced in the 107th Congress. The comprehensive energy bill adopted by the U.S. Senate in 2002 requires steady increases in the amount of ethanol that is mixed with gasoline until it reaches 5 billion gallons per year in 2012. A 50 percent volumetric bonus is provided for bioethanol produced from cellulosic materials instead of corn.

This policy would establish a CO_2 emissions standard for gasoline, starting at a 5 percent reduction relative to the current average emissions factor by 2010 and increasing 1 percent per year to a 15 percent reduction by 2020. The standard would be complemented by expanded R&D, market creation programs, and financial incentives to stimulate the production and demand for low-carbon fuels such as bioethanol and biomass- or solar-based hydrogen.

It is assumed that most of the reduction in average carbon content is provided by bioethanol. Bioethanol, unlike ethanol made from corn, is produced from low-cost agricultural residues, forestry residues, or dedicated herbaceous or woody energy crops. Bioethanol offers the potential for both lower costs than corn-based ethanol and near-zero or negative CO_2 emissions (Interlaboratory Working Group 2000).[7] Bioethanol production techniques are under active development, but uncertainties regarding the performance of full-scale plants, their financial viability, and market potential are holding back commercialization (CEC 2001a, Wyman 1999). In addition to bioethanol, ethanol could be produced from sugarcane using techniques developed and refined in Brazil.

Most of the ethanol resulting from this policy is likely to be blended into gasoline, although some could be used in vehicles designed to operate on pure or nearly pure ethanol. The latter requires development of an ethanol fuel delivery infrastructure, however. Also, it is assumed that the cost of producing bioethanol declines from about $1.40 per gallon today to $0.91 per gallon by 2020 (in 1999 dollars) as a result of continued R&D as well as technology learning and economies of scale as production grows (Interlaboratory Working Group 2000).

Energy, Economic, and Environmental Impacts

The Clean Energy Scenario is compared to a business-as-usual Base Scenario to analyze its potential energy, economic, and environmental impacts.[8] All 10 policies in the Clean Energy Scenario are analyzed together to avoid overlap and double counting of savings. In addition, each policy is analyzed individually to estimate its own contribution to savings or supply. In doing so, the energy savings policies were examined first, and then the energy supply policies were considered.[9]

The Base Scenario is derived from and very similar to the Reference Case forecast in the *Annual Energy Outlook 2001*. Several changes were made to the EIA's assumptions regarding renewable energy technologies to better reflect their costs and prospects. Also, the passenger vehicle fuel economy improvement assumptions included in the EIA Reference Case were modified. As noted previously, new passenger vehicle fuel economy has stagnated or declined for the past 15 years. EIA assumes this trend will suddenly reverse and that passenger vehicle fuel economy will increase by 16 percent between 2000 and 2020 in the absence of new policies (EIA 2000c). This assumption is unrealistic, so vehicle fuel economy improvement was scaled back in the Base Scenario.

All fuel prices in the Base Scenario are derived from the *Annual Energy Outlook 2001*. Energy prices in the Clean Energy Scenario are modeled taking into account the demand reductions and shifts in energy supply mix caused by the policies. Economic growth is assumed to be the same in the Base and Clean Energy Scenarios—3 percent per year on average between 1999 and 2020 (EIA 2000c). The costs of efficiency measures are amortized over the life of each measure to account for costs and benefits in a consistent manner.[10] For efficiency measures with a life that extends beyond 2020, not all costs or benefits are included in the analysis. Ultimately, the analysis paints two very different pictures of what our energy future might look like in terms of energy sources, demand growth, costs, and environmental impacts.

Table 5-2 summarizes the overall energy use, carbon and other pollutant emissions, and economic impacts for 2010 and 2020 in the two scenarios. In the Base Scenario, total primary energy consumption reaches nearly 115 quads by 2010 and 128 quads by 2020, a 1.3 percent annual average growth rate. In the Clean Energy Scenario, the 10 policies reduce primary energy consumption 11 percent by 2010 and 26 percent by 2020, relative to energy use in the Base Scenario. If all the policies are adopted, primary energy use would rise slightly during the next decade but fall between 2010 and 2020 (Fig. 5-3). By 2020, primary energy use would be slightly lower than the level in 1999, even though population is assumed to rise by 19 percent and GDP by 86 percent over this period (EIA 2000c).

Renewable energy use (other than hydropower) would grow from 3.4 quads in 1999 to about 9.2 quads by 2010 and 12 quads by 2020 in the Clean Energy Scenario. This overall growth rate, 6 percent per year on average, is three times faster than the renewable energy growth rate in the Base Scenario. Renewable energy

TABLE 5-2

Comparison of the Base and Clean Energy Scenarios

	1999	2010 Base Scenario	2010 Clean Energy Scenario	2020 Base Scenario	2020 Clean Energy Scenario
End use energy (quads)	71.6	86.5	79.2	98.3	78.4
Primary energy (quads)	96.1	114.6	103.1	128.1	94.4
Coal	21.4	25.2	17.3	26.2	5.7
Oil	38.0	44.9	40.6	51.7	39.3
Gas	22.0	28.7	24.8	35.5	28.2
Nuclear	7.8	7.7	7.8	6.1	5.9
Hydro	3.2	3.1	3.1	3.1	3.1
Nonhydro renewables	3.4	4.8	9.2	5.3	12.0
Carbon emissions (million metric tons)	1,505	1,817	1,479	2,063	1,202
Other emissions (million metric tons)					
Sulfur dioxide	15.75	12.79	11.52	12.68	4.90
Nitrogen oxides	20.51	16.48	15.35	16.88	13.11
Particulate matter (PM-10)	1.45	1.49	1.38	1.63	1.36
Cumulative net economic savings ($1999 billions)		—	146	—	554
U.S. gross domestic product (billion 1999$)	9,273	13,234	13,234	17,254	17,254
Energy intensity (kBtu /1999$)	10.4	8.7	7.8	7.4	5.5

including hydropower would provide 16 percent of U.S. primary energy supplies in 2020 in the Clean Energy Scenario, compared to about 7 percent in 1999.

Oil use increases by about one-third by 2020 in the Base Scenario, with oil imports increasing more than 60 percent over this time period. Oil imports would rise from about 55 percent of total U.S. oil consumption today to about 70 percent of total consumption by 2020. The policies in the Clean Energy Scenario would significantly reduce the growth in oil imports. Relative to the Base Scenario, oil use by 2010 is about 10 percent lower, whereas annual imports decrease by about 14 percent, assuming that domestic oil production is the same in both scenarios. By 2020, oil use is about 24 percent and imports about 35 percent lower in the Clean Energy Scenario, relative to levels in the Base Scenario.

Coal consumption increases by 22 percent between 1999 and 2020 in the Base Scenario. In contrast, coal use declines 73 percent over this period in the Clean Energy Scenario due to efficiency improvements and the replacement of many coal-fired power plants with natural gas and renewable energy–based power

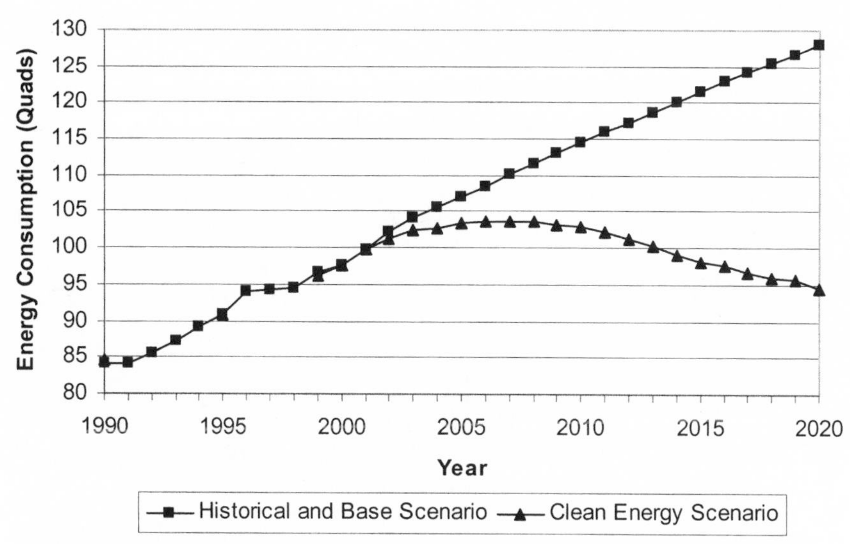

FIGURE 5-3: Primary energy consumption in the Base and Clean Energy Scenarios.

sources. Thus the Clean Energy Scenario envisions a large-scale shift away from coal in the United States, similar to the shift already made in the United Kingdom in a shorter time frame (see Chapter 4).

Because of increased use of natural gas for electricity generation, natural gas use grows 29 percent between 1999 and 2020 in the Clean Energy Scenario, indicating that some increase in natural gas supplies will be needed. But the growth in natural gas use in the Clean Energy Scenario is substantially lower than the 62 percent increase in the Base Scenario.

Electricity use in 2020 would be about the same as in 1999 in the Clean Energy Scenario although growth in renewable sources and CHP systems would greatly decrease the need for fossil fuel–based and nuclear power. All renewable sources (including hydro power) would provide about 21 percent of total electric supply in 2010 and 26 percent in 2020 in the Clean Energy Scenario. CHP systems provide 13 percent of electricity supply in 2010 and nearly 31 percent in 2020. For comparison, renewable energy sources and CHP systems together provided about 16 percent of electricity supply as of 1999. The percentage of electricity produced by these options does not increase in the Base Scenario.

National energy intensity continues to decline about 1.6 percent per year on average between 1999 and 2020 in the Base Scenario (Fig. 5-4). This reduction is caused by structural shifts in the economy and energy savings from energy efficiency standards already adopted; ongoing federal, state, and local efficiency programs; and anticipated technological innovation. In the Clean Energy Scenario,

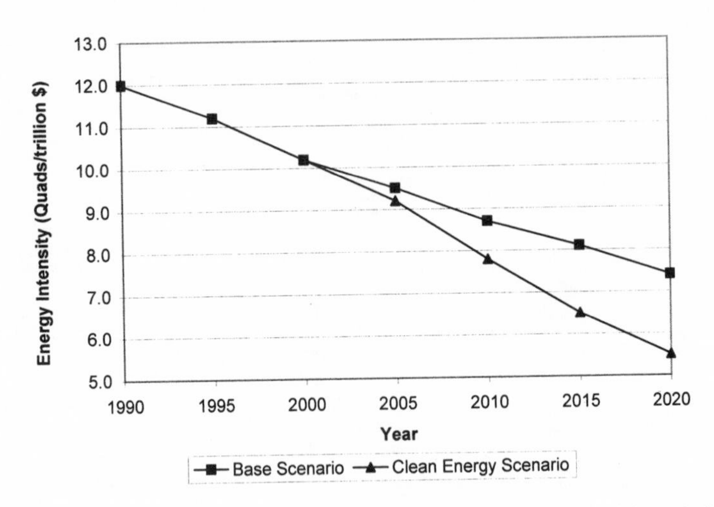

FIGURE 5-4: Energy intensity in the Base and Clean Energy Scenarios and historical values.

national energy intensity falls 3.0 percent per year on average through 2020, nearly twice the rate in the Base Scenario. A 3 percent rate of decline in energy intensity is slightly more than the rate experienced from 1973 to 1986 but slightly less than the rate from 1996 to 2000. Thus the policies have a significant, but not extreme, impact on energy growth and energy intensity.

Figure 5-5 summarizes the direct economic costs and benefits of the 10 policies in the Clean Energy Scenario. The policies induce incremental investments in more efficient buildings and appliances, more fuel-efficient cars and trucks, cleaner and more efficient power plants, renewable energy sources, and so on. The total investment in efficiency measures equals about $487 billion and the investment in renewable technologies about $186 billion through 2020.[11] To place these figures in context, the annual energy bill (i.e., all retail energy purchases) equaled about $558 billion as of 1999 (EIA 2000c).

The implementation of energy efficiency and renewable energy measures leads to less conventional energy use as well as some operating-cost savings in areas such as petroleum refining. Overall, consumers and businesses will save about $1.2 trillion dollars through 2020 as a result of these policies. Thus energy bill and operating savings more than offset the investment costs, with net savings of about $145 billion through 2010 and $554 billion through 2020. The net savings grow over time since energy efficiency and renewable energy measures have more time to pay back their initial cost.

Table 5-3 shows the cost-effectiveness of each policy, considering all costs

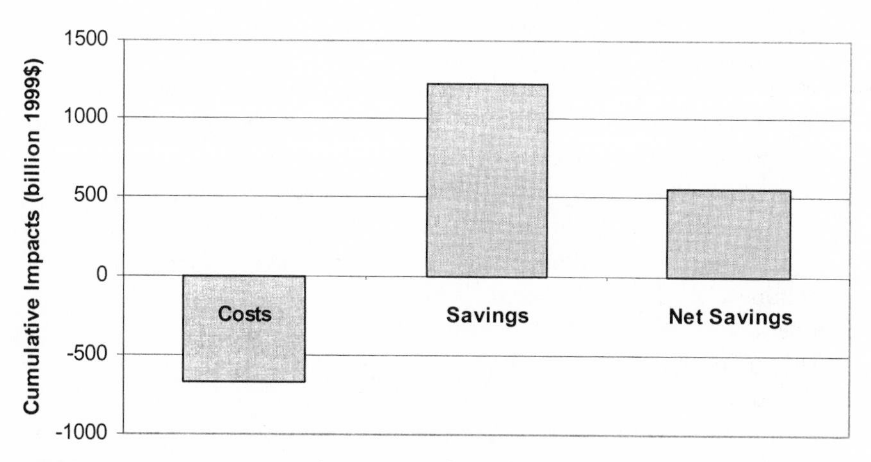

FIGURE 5-5: Cost, savings, and net savings from the policies (cumulative present value through 2020).

TABLE 5-3

Economic Costs and Benefits by Policy (cumulative present value by 2020, billion 1999$)

	Costs	Savings	Net Savings
Increase passenger vehicle fuel economy standards	102	251	148
Establish a national system benefits trust fund	130	231	101
Adopt voluntary agreements to reduce industrial energy use	48	159	112
Establish a renewable portfolio standard for power generators	44	26	–19
Adopt new appliance efficiency standards and stronger building codes	37	145	107
Provide tax incentives for innovative renewable energy and energy-efficient technologies	17	26	8
Expand federal R&D and deployment programs	33	86	53
Remove barriers to combined heat and power systems	63	189	125
Establish renewable energy or carbon content standards for vehicle fuel	142	123	–19
Strengthen emissions standards on coal-fired power plants	57	–7	–64
TOTAL CLEAN ENERGY SCENARIO	674	1229	554

and savings through 2020. The demand-side policies in aggregate are very cost-effective, with savings that are nearly three times their costs, thereby yielding net benefits of about $655 billion. On the other hand, the supply-side policies are not cost-effective based on their direct costs and benefits (i.e., ignoring environmental and social impacts). The two renewable energy policies and coal-fired power plant policy, in combination, have investment costs that exceed savings by $102 billion. Although not every policy is cost-effective on its own, combining all of the policies results in a net savings of $554 billion over the 20-year period.

The 10 policies would also have a positive economic impact by reducing the price of some forms of energy as a consequence of lower energy demand. In the Clean Energy Scenario, natural gas prices are projected to decline by 37 percent relative to prices in the Base Scenario by 2020, and are expected to fall below 1999 levels. Coal prices would also fall due to dramatically lower coal use in the Clean Energy Scenario.[12] But electricity prices are slightly higher in the Clean Energy Scenario relative to the Base Scenario due to the RPS and tighter emissions standards.

Figure 5-6 presents energy expenditures for the Base and Clean Energy Scenarios incorporating both the direct and the indirect price effects. In the Base Scenario, U.S. energy expenditures are projected to rise from $557 billion in 1999

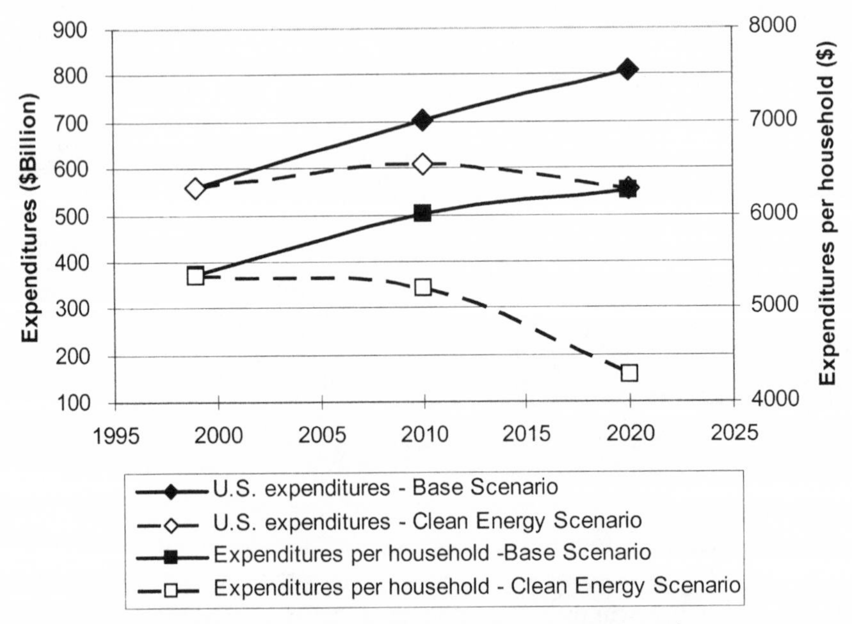

FIGURE 5-6: Energy expenditures in the Base and Clean Energy Scenarios.

to $809 billion in 2020 (in 1999 dollars). In the Clean Energy Scenario, total energy expenditures would remain steady and be at $555 billion in 2020, a 31 percent savings relative to estimated expenditures that year in the Base Scenario. Energy expenditures per household (including energy used in homes, transportation, and business) increase from $5,355 in 1999 to $6,249 in 2020 in the Base Scenario. In the Clean Energy Scenario, energy expenditures per household drop to $4,287 in 2020.

The Clean Energy Scenario should also have a positive effect on employment and economic growth. Although these impacts were not explicitly analyzed for this set of policies, other studies have determined that increasing energy efficiency and expanding renewable energy use would result in a net increase in employment (Bernow et al. 1999, Laitner, Bernow, and DeCicco 1998). This issue is further examined in Box 5-1.

U.S. CO_2 emissions in the Base and Clean Energy Scenarios are shown in

BOX 5-1

Clean Energy Development and Jobs

The transition to a cleaner, more sustainable energy future, like all economic transitions, will involve shifts in employment. There will be job losses in the fossil fuel and nuclear energy industries, but job gains in manufacturing, installing, and servicing energy efficiency and renewable energy technologies. The net impact will be positive—more jobs will be created than lost as this transition occurs.

The net employment gains result because conventional energy supply—producing oil and natural gas, mining coal, and generating and distributing electricity—is not very labor intensive. Increasing energy efficiency and supplying renewable energy are much more labor intensive than conventional energy supply. Also, the money that consumers and businesses save as a result of energy efficiency improvements gets respent throughout the economy in ways that add jobs overall (Laitner, Bernow, and DeCicco 1998).

The number of jobs in the fossil fuel industries is relatively low and declining as labor productivity in these industries continues to rise. In the United States, coal mining employment fell from 243,000 workers in 1980 to just 83,000 workers in 1999 in spite of increasing coal production (Renner 2000). This steep drop was due to mechanization and the ongoing shift from underground to surface mining. Coal mining employment has also fallen dramatically in the United Kingdom, Germany, China, and elsewhere.

Employment in the oil, gas, and electric utility industries has also declined in the United States and other countries in the past 20 years in spite of increasing output (Renner 2000).

Employment in the renewable energy industries, on the other hand, is rising. The wind power industry supported about 15,000 jobs in Germany as of 1998 and about 20,000 jobs in Denmark as of 2001 (Krohn 2002b, Renner 2000). One study estimates that if the wind industry continues to grow about 25 percent per year, it could provide 10 percent of electricity consumed worldwide and support 1.7 million jobs by 2020 (EWEA and Greenpeace 1999). Likewise, if the solar PV industry continues to grow 25 to 30 percent per year, it could support 2 million jobs worldwide by 2020 (Cameron et al. 2001).

The jobs created as energy efficiency improves and renewable energy use expands would be widely dispersed geographically, unlike employment in the fossil fuel industries. Many of these jobs would be skilled positions with good pay. Jobs and income would flow to both urban and rural areas; to the former for manufacturing and installing these devices, and to the latter for growing energy crops or locating wind turbines on farms or open land. And the energy efficiency and renewable energy industries, unlike coal mining and oil and gas production, would be relatively safe for workers and those living near where the energy is produced.

Various studies have analyzed the net impact on jobs from a broad set of initiatives to increase energy efficiency and renewable energy use in particular countries, often in the context of considering strategies to reduce carbon dioxide emissions. Table 5-4 shows the results of studies done for six countries. In all cases, there would be a net employment gain at the same time carbon emissions fall. The net gain in the United States, for example, is equivalent to a roughly 0.5 percent reduction in the unemployment rate.

Efforts should be made to help workers who lose their jobs in the transition to a clean energy future. They should be retrained, possibly using a portion of funds collected through new energy or carbon dioxide emissions taxes (if such taxes are adopted). Also, renewable energy industries should be fostered in regions particularly hard hit during the transition. In the United States, special efforts could be made to develop bioenergy and wind power industries in regions that lose coal mining jobs such as Appalachia or the western high plains.

◉ ◉ ◉

TABLE 5-4

Estimated Job Impacts from Clean Energy Strategies in Six Countries

Country	Strategy	Time Period	Carbon Emissions Reduction (MMT)	Employment Change (net jobs)
Austria	Cogeneration, energy efficiency, renewable energy, alternative transportation, higher fossil fuel taxes	1997–2005	70	+42,000
Denmark	Cogeneration, district heating, energy efficiency, renewable energy use	1996–	82	+16,000
Germany	Energy efficiency, renewable energy use, nuclear power phase-out, alternative transportation	1990–2020	518	+208,000
Netherlands	Energy efficiency, wind power	1995–2005	440	+71,000
United States	Energy efficiency, Renewable energy use, cogeneration, alternative transportation	1995–2010	4,200	+870,000

SOURCES: Renner 2000, Bernow et al. 1999.

Figure 5-7. In the Base Scenario, CO_2 emissions from burning fossil fuels reach 1,817 MMT of carbon by 2010 and 2,062 MMT by 2020, meaning a 1.5 percent annual average growth rate during 2000–2020. Relative to emissions in 1990, emissions are 36 percent greater by 2010 and 54 percent greater by 2020. In the Clean Energy Scenario, CO_2 emissions fall to 1,479 MMT of carbon by 2010 and 1,202 MMT by 2020. This level of reduction would not be adequate to reach the United States's Kyoto Protocol target of holding emissions to 7 percent below 1990 emissions between 2008 and 2012, but it would be a major step in that direction.[13]

If the policies in the Clean Energy Scenario were adopted, the United States should be able to achieve its Kyoto Protocol target through some combination of (1) further efforts to increase energy efficiency and renewable energy use (e.g., by adopting taxes on energy or emissions, taking actions to reduce car and light truck use, or increasing the efficiency of freight transport); (2) larger reductions in emissions of other greenhouse gases covered under the Kyoto Protocol; (3)

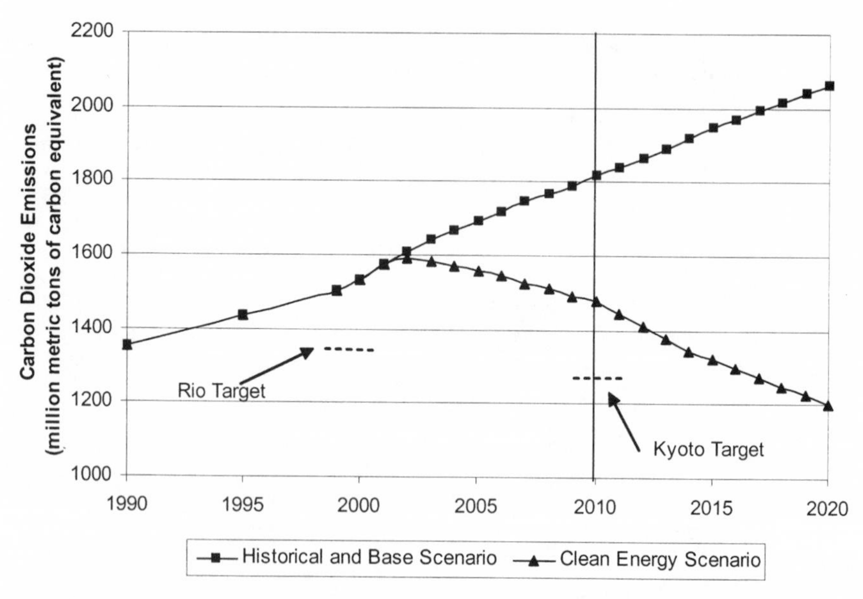

FIGURE 5-7: Carbon dioxide emissions in the Base and Clean Energy Scenarios.

purchase of emissions reductions from other countries and from Clean Development Mechanism projects; and (4) carbon sequestration through forestation efforts as permitted under the Protocol. A recent study for the World Wildlife Fund examines the impacts of these other actions, in combination with the policies recommended here, and concludes that the United States could meet its Kyoto target without penalizing the U.S. economy (Bailie et al. 2001).

Carbon emissions continue to fall after 2010 in the Clean Energy Scenario as adoption of energy efficiency and renewable energy measures expands. Compared to the Base Scenario, carbon emissions are cut 864 MMT (42 percent) by 2020. Relative to current emissions levels, U.S. energy-related carbon emissions decline about 23 percent by 2020 as a result of the 10 policies. This level of carbon emissions reduction is consistent with a climate stabilization scenario whereby industrialized nations cut their absolute carbon emissions over 50 percent by 2050 and 90 percent or more by 2100 (Bailie et al. 2001, PCAST 1997).

Table 5-5 shows the carbon emissions reductions from each of the 10 policies. Stronger CAFE standards, industrial voluntary agreements, and the national system benefits trust fund are the policies that yield the largest reductions. It may appear surprising that tighter emissions standards on coal-fired power plants cause rather modest carbon reductions of 43 MMT in 2010

TABLE 5-5

Carbon Emissions Reductions by Policy (MMT/year)

	2010	2020
Total Base Scenario carbon emissions	1817	2063
Increase passenger vehicle fuel economy standards	40	142
Establish a national system benefits trust fund	46	127
Adopt voluntary agreements to reduce industrial energy use	67	132
Establish a renewable portfolio standard for power generators	36	81
Adopt new appliance efficiency standards and stronger building codes	29	99
Provide tax incentives for innovative renewable energy and energy-efficient technologies	4	10
Expand federal R&D and deployment programs	19	65
Remove barriers to combined heat and power systems	29	78
Establish renewable energy or carbon content standards for vehicle fuel	25	55
Strengthen emissions standards on coal-fired power plants	43	71
Total Clean Energy Scenario carbon emissions	1479	1202

NOTE: MMT/year = million metric tons per year.

and 71 MMT in 2020. This is because the other policies result in significant reductions in conventional electricity generation, and carbon reductions from the supply-oriented policies are computed after the reductions from the demand-oriented policies.

The set of 10 policies reduces other pollutant emissions besides CO_2. Implementing the policies cuts SO_2 emissions the most—61 percent by 2020, relative to emissions in the Base Scenario (see Fig. 5-8). Emissions of NO_x are cut 22 percent and particulate emissions drop about 17 percent by 2020. These emissions reductions would result in significant air quality and public health benefits (ALA 2001, Clean Air Task Force 2000).

The Clean Energy Scenario would also result in other environmental and public health benefits by reducing coal mining and pressure to expand oil and natural gas production in the United States. Coal mining is a very hazardous occupation (Holdren and Smith 2000). It also disrupts the land, generates large quantities of solid waste, and pollutes water supplies (NRDC 2001). By significantly cutting oil and natural gas use relative to the Base Scenario, the Clean Energy Scenario is compatible with keeping the pristine Arctic National Wildlife Refuge as well as other environmentally sensitive areas such as currently protected portions of the Outer Continental Shelf and western wilderness areas off-limits for oil and gas drilling.

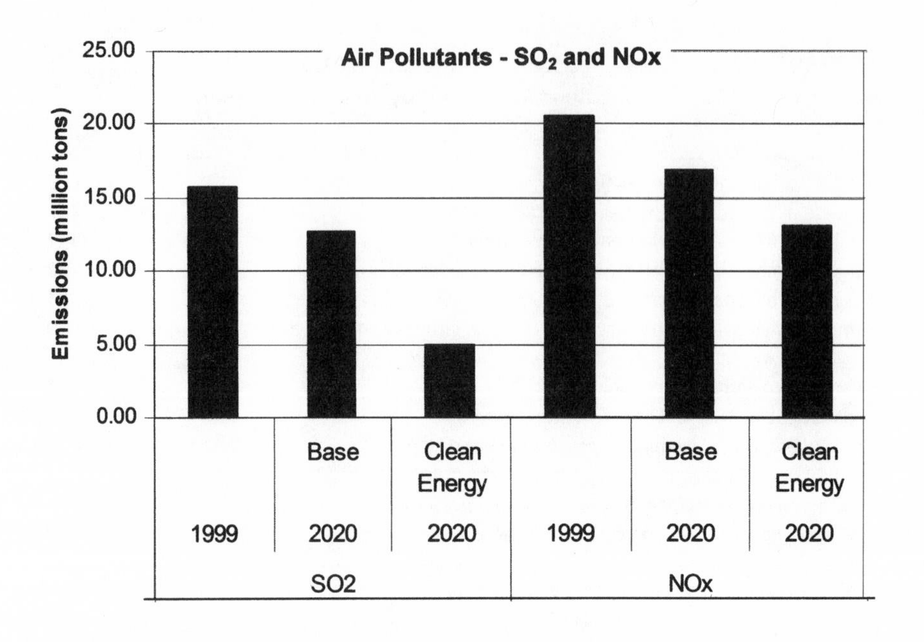

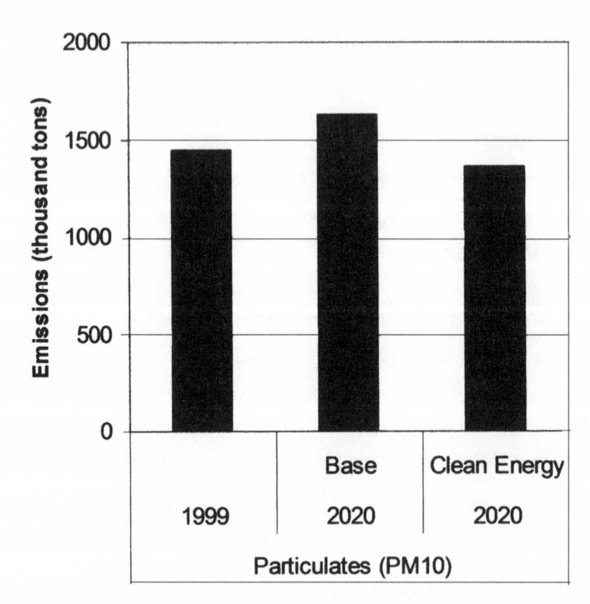

FIGURE 5-8: Criteria pollutant emissions in the Base and Clean Energy Scenarios.

Summary

This analysis confirms that energy efficiency improvement and renewable energy deployment can and should be the cornerstone of America's energy policy. Aggressively pursuing energy efficiency and renewable energy sources, as envisioned in the Clean Energy Scenario, would:

- reduce oil imports thereby enhancing national security;
- save consumers and businesses hundreds of billions of dollars;
- significantly cut U.S. CO_2 emissions relative to current trends, enabling the United States to get on track for deep emissions reductions over the long run;
- reduce SO_2, NO_x, and particulate emissions, thereby improving public health; and
- reduce the need for fossil fuel development in environmentally sensitive areas.

In short, the 10 policies would place the United States on a path to a more sustainable energy future. But a comprehensive and strong set of policies is needed to make this transition. Watering down or dropping some of the policies would reduce their overall impacts and benefits. The high level of energy savings and increased renewable energy use will not occur on their own due to the many barriers described in Chapter 2.

The transition to a more sustainable energy future in the United States is not inhibited by a lack of technologies or lack of policy mechanisms. The technologies assumed in the Clean Energy Scenario are either commercially available today or emerging in the marketplace. Most of the recommended policies have been successfully adopted either at the state level or at the national level previously, and a few have been successfully adopted outside the United States. California, for example, enacted a number of the policies recommended here and cut its energy intensity and carbon emissions intensity much faster than the United States as a whole over the past 25 years.

There is also strong public support for the transition to a more sustainable energy future. Citizens consistently support increasing energy efficiency and expanding renewable energy use over drilling for more oil and natural gas or building new coal-fired or nuclear power plants. Recent Gallup polls, for example, show that 90 percent of the public favor investing in renewable energy sources and about 80 percent favor requiring new vehicles to have higher fuel economy (Gillespie 2001). A majority of the public, on the other hand, opposes opening the Arctic Refuge to oil drilling or increasing the use of nuclear power.

The main obstacles to a more sustainable energy future, as envisioned in the

Clean Energy Scenario, are inertia, opposition from powerful industries, and lack of political will. As noted in Chapter 2, the auto industry strongly opposes raising the fuel economy standards on new vehicles. The utility industry generally opposes a renewable portfolio standard, a national systems benefit charge, or policies that would accelerate the retirement of older coal-fired power plants. Many manufacturers and builders oppose new appliance efficiency standards or stronger building energy codes. And the oil industry opposes a mandate for increasing the supply of renewable-based fuels.

These industries have a tremendous influence on national energy policy through their campaign contributions, presence in Washington, D.C., and sheer size. Oil and gas companies and utilities contributed heavily to the Bush–Cheney campaign as well as the campaigns of many members of Congress (Box 5-2). The Bush–Cheney energy plan as well as House of Representatives plan backs policies these companies favor, such as allowing oil drilling in the Arctic Refuge, more oil and gas production on public lands, construction of additional nuclear and coal-fired power plants, weaker environmental standards on energy facilities, and no restrictions on CO_2 emissions. And neither Democratic nor Republican administrations have been willing to take on the auto lobby and raise vehicle fuel economy standards.

BOX 5-2

Campaign Contributions and Energy Policy

The energy and automotive industries are major contributors to political campaigns in the United States. In the 2000 election cycle (1999–2000), energy and natural resources companies contributed $65 million to congressional and presidential campaigns. Three-quarters of this went to Republicans and one-quarter to Democrats. The single largest contributor was the Enron Corporation, which gave $2.4 million to political campaigns and strongly advocated deregulation prior to its collapse. The second-largest contributor was the Southern Company, a major utility holding company, which gave $1.4 million to political campaigns in this cycle. The Southern Company has long fought tougher emissions standards on power plants. The third-largest contributor was the ExxonMobil Corporation, which also gave about $1.4 million. ExxonMobil strongly opposes renewable fuels requirements, the Kyoto Protocol, or any limits on the emissions of CO_2 and other heat-trapping gases. For comparison, all renewable energy companies combined contributed about $800,000 to political campaigns in this election cycle.

The automotive industry contributed over $18 million to political campaigns in the 1999–2000 election cycle. This includes contributions from auto dealers as well as auto manufacturers. Nearly 80 percent of their contributions went to Republican candidates and 20 percent to Democrats. The automotive industry strongly opposes raising the fuel economy standards on new cars and light trucks. Both Congress and the Executive branch have failed to raise these long out-of-date standards, and proposals to raise the standards were defeated once again in the Senate in early 2002.

The Enron debacle and subsequent revelations about it demonstrate that campaign contributions buy access and influence in Washington, D.C. Enron and its former chairman Ken Lay were the largest single contributor to the Bush political dynasty. Not surprisingly, Enron influenced the Bush–Cheney Energy Plan as well as Bush administration appointments to the Federal Energy Regulatory Commission and other agencies. Enron also heavily contributed to the campaigns of key members of Congress—members who supported energy and securities deregulation that enabled Enron to create its complex and deceitful trading schemes.

SOURCE: Center for Responsive Politics 2001, Phillips 2002.

◉ ◉ ◉

While it is difficult for national policy makers in the United States to resist pressure from the fossil fuel and auto industries, policy makers in other countries, including western European countries and Japan, are adopting relatively strong policies to increase energy efficiency and renewable use. Likewise, some developing countries have adopted successful initiatives to cut their energy waste and expand renewable energy use, as noted in the case studies pertaining to Brazil, China, and India in Chapter 4. But, like industrialized countries, developing countries could further benefit from comprehensive policies to increase energy efficiency and renewable energy use. Chapter 6 illustrates this opportunity for a key developing country—Brazil.

Notes

1. These figures are based on commercial energy consumption and exclude fuelwood and other traditional, noncommercial energy sources.
2. Passenger vehicles include cars, SUVs, minivans, and pickup trucks.

3. This estimate corrects for inflation and values GDP in constant 1990 dollars.
4. The Energy Information Administration is the data collection and energy forecasting branch of the U.S. Department of Energy.
5. These figures, and savings values cited for other policies recommended here, come from the "Energy, Economic, and Environmental Impacts" section later in this chapter.
6. Increased funding would be used both for R&D and for education training, as well as other deployment-oriented activities.
7. Negative CO_2 emissions can result if ethanol and electricity are coproduced and the avoided emissions from power plants are greater than the emissions from producing and processing the biomass.
8. The DOE/EIA National Energy Modeling System (NEMS) is used to estimate these impacts. NEMS is a computer model that projects future energy consumption and supply based on energy technology and fuel choices for each sector and end-use. It takes into account technology characteristics, equipment turnover rates, and financial and behavioral parameters (EIA 2000e).
9. This analysis was conducted by the Tellus Institute, Boston, MA. It is an expanded version of a study performed by the American Council for an Energy-Efficient Economy (ACEEE) and the Tellus Institute (Nadel and Geller 2001).
10. A 5 percent real discount rate is assumed in the cost–benefit analysis. The discount rate reduces future costs and benefits due to the time value of money.
11. All costs are expressed in 1999 dollars.
12. The "feedback effect" of lower energy prices on energy demand is incorporated into the analysis of the *Clean Energy Scenario.*
13. The Kyoto Protocol target applies to six greenhouse gases, of which CO_2 is by far the most important (see Table 1-1). For energy-related CO_2 emissions, the U.S. target of 7 percent below 1990 emissions is equal to 1,257 MMT of carbon equivalent.

Brazil: Policies and Scenarios

About 4.9 billion people, 80 percent of the world's population, lived in developing countries as of 2001. And population is now growing about 1.5 percent per year in developing countries, compared to just 0.2 percent per year in the industrialized nations (UNFPA 2001). But per capita energy consumption is much lower in developing countries compared to industrialized nations. Developing countries account for 39 percent of global energy consumption and only 32 percent of the global consumption of modern energy sources (IEA 2000a). Given the low living standards and low per capita energy use in developing countries (including approximately 2 billion people consuming little or no modern energy sources), total energy use in developing countries is increasing fairly rapidly. Most of this growth is expected to come from oil and coal in a business-as-usual energy scenario at least for the next two decades (IEA 2000a).

Developing countries need to increase their energy consumption to fuel their social and economic development. But the energy resources and technologies they choose, and the distribution of these resources and technologies among their populations, will affect future living conditions in these countries. These choices will also have a dramatic effect on whether the world limits the risk of global warming or allows it accelerate out of control.

There are great differences among developing countries with respect to size, social and economic conditions, and patterns of energy use. Impoverished nations in Africa and Asia are highly dependent on traditional energy sources, while wealthier developing nations have largely made the transition from traditional to modern energy sources. This chapter explores the policy options for advancing a cleaner and more sustainable energy future in one of the wealthier developing countries—Brazil. It shows that policy choices can have a significant impact on energy trends, social progress, and environmental quality in a developing country.

Brazil is the fifth most populous country with roughly 172 million inhabitants. Brazil is also the ninth largest economy in the world with a gross domestic product (GDP) per capita of approximately US$3,300 as of 2000. This makes Brazil a middle-income developing country. It is the largest country in Latin America in terms of economy, population, and land area. Brazil is highly urbanized with nearly 80 percent of the population living in urban areas and about 10 percent of its population living in metropolitan São Paulo. Brazil also has a very inequitable income distribution and high poverty rates in some regions. Poverty is most acute in northeastern Brazil and in rural areas, where 50 percent or more of families earn $150 per month or less[1] (PNAD 1999).

Energy use in Brazil increased rapidly over the past 25 years (Fig. 6-1). Total energy use increased by nearly 250 percent between 1975 and 2000, while energy use per capita increased 60 percent and energy use per unit of GDP increased 22 percent. Rapid industrialization, including high growth of some energy-intensive industries such as aluminum and steel production, and increasing residential and commercial energy services were the main causes of this growing energy use and energy intensity (Tolmasquim et al. 1998).

Figure 6-1 also shows the evolution of different energy sources over the past 25 years. Brazil greatly expanded its park of hydroelectric plants, with total hydro capacity increasing from 16 gigawatts (GW) in 1975 to 60 GW in 2000.[2] Hydropower provided 90 percent of all electricity and 39 percent of all energy consumed in Brazil as of 2000.[3] Petroleum is the second-largest source of energy, accounting for 34 percent of total energy use as of 2000. Fuel substitution and conservation efforts, including the national ethanol fuel program, limited growth in consumption of petroleum products over the past 25 years. Coal and natural gas are minor sources of energy in Brazil, although natural gas supply is now increasing rapidly.

Bioenergy sources, including wood, charcoal, and sugarcane products (ethanol and bagasse) provided about 16 percent of energy consumed as of 2000.

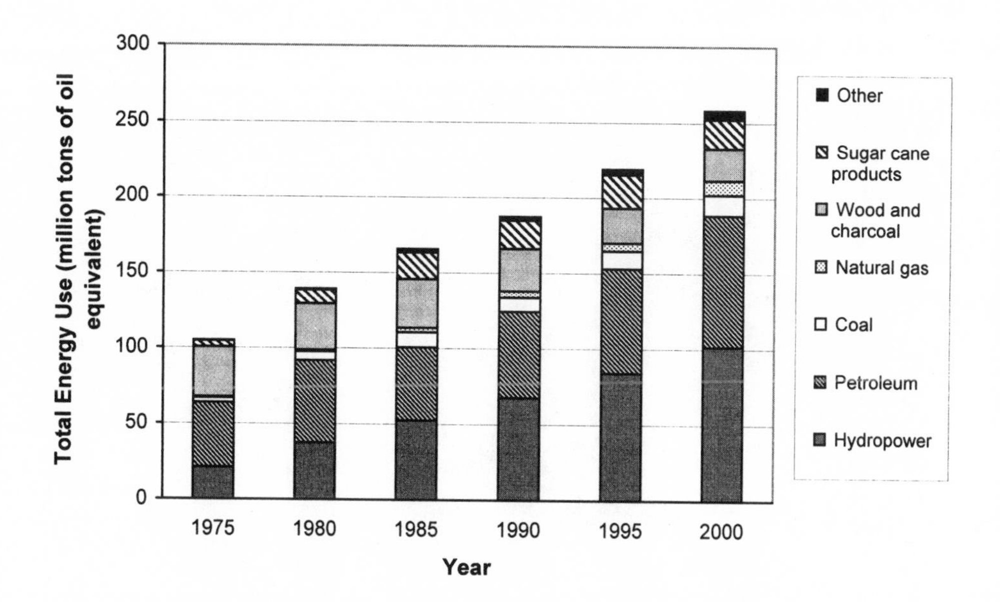

SOURCE: MME 2000.

FIGURE 6-1: Trends in primary energy use in Brazil.

Consumption of fuelwood and charcoal declined by one-third over the past 25 years, but was compensated for by the growth in sugarcane products as an energy source. With its high dependence on hydroelectricity and bioenergy, renewable energy sources accounted for about 56 percent of total energy supply as of 2000. This is a very high renewable energy fraction for a middle-income developing nation. Furthermore, fuelwood represented only about 5 percent of all energy and less than 10 percent of the renewable energy total in 2000. For reference, fuelwood provided about 76 percent of all energy consumed in Brazil in 1941 (Oliveira et al. 1998).

Figure 6-2 shows the consumption of energy by sector in 1975 and 2000. The industrial sector is the largest energy consumer and increased its share of total energy consumption over the past 25 years. The commercial and energy sectors also increased their shares, while the residential and transportation shares declined.[4] The steep reduction in the residential share is due to the displacement of fuelwood by modern, more efficient energy sources. Fuelwood accounted for 74 percent of residential energy use in 1975 compared to just 17 percent in 2000. At the same time, electricity use increased from 16 percent of the residential total in

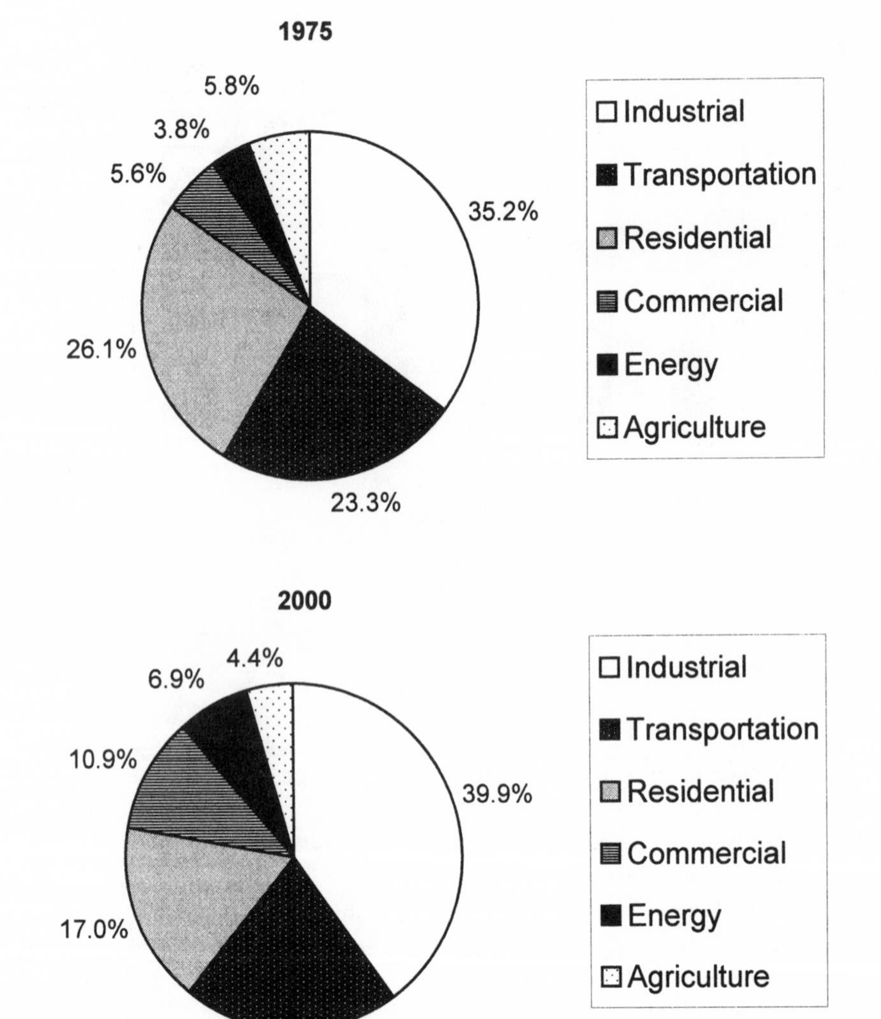

SOURCE: MME 2000.

FIGURE 6-2: Final energy consumption by sector.

1975 to 64 percent in 2000. In addition, bottled gas use for cooking greatly expanded, accounting for 17 percent of total residential energy consumption in 2000 (MME 2000).

Energy policy in Brazil over the past 25 years mainly attempted to reduce the country's dependence on foreign energy supplies and stimulate the development of domestic energy sources. Policies were devised to increase domestic oil production, expand alcohol fuel production and use, generate nuclear energy, and conserve energy. The positive experiences with alcohol fuel production and electricity conservation were described in Chapter 4. Efforts to expand domestic oil output, including developing new techniques for oil production in deep waters, were also very successful. Domestic oil production increased from about 0.2 million barrels per day in 1980 to nearly 1.4 million barrels per day in 2000–01. These policies and their outcomes benefited the country's balance of trade, national security, capital goods industry, and labor market.

During the 1990s, energy policy concentrated on privatizing and restructuring both the petroleum and the power sectors. Also, an effort was made to stimulate the development and utilization of natural gas in Brazil. These initiatives have had mixed success so far. Electricity sector privatization and restructuring is in midstream. Flaws in this strategy led to relatively little investment in new generation and transmission facilities during the late 1990s, which in turn contributed to a severe power shortage in 2001 (Tolmasquim 2001). The supply of natural gas is increasing, but demand has not matched supply due in large part to the high cost of gas imports.

Overall, Brazil has successfully implemented some but not all energy policies over the past 25 years. Policies for increasing modern renewable energy sources and domestic petroleum supply were very successful. Policies for increasing energy efficiency and expanding natural gas use met with limited success. Nonetheless, a variety of new energy policies and initiatives could help Brazil to advance socially and economically, as well as achieve other important objectives.

Objectives

A developing country like Brazil has some of the same objectives and interests as the United States and other industrialized nations with respect to energy policy—namely to diversify its energy sources, reduce its import dependence, and cut inefficiency and energy waste. But Brazil and other developing countries also have some differing objectives and priorities—namely to ensure adequate energy supplies and avoid energy shortages, limit investment requirements for meeting energy service needs, and foster social development. The range of energy policy objectives in Brazil is briefly reviewed in the following text.[5]

Diversify Energy Supplies

As already noted, energy supply in Brazil is dominated by two forms of energy: petroleum and hydroelectricity. Heavy dependence on petroleum leaves Brazil vulnerable to price spikes and "shocks" since prices for domestic petroleum mirror those in the world market. Heavy dependence on hydropower leaves Brazil vulnerable to electricity shortages due to periodic droughts. As previously mentioned, Brazil experienced a power shortage in 2001 due to inadequate investment in power generation and transmission in recent years, coupled with drought conditions that reduced hydropower output. Consumers in most of the country were required to reduce their electricity use by 20 percent. Failure to comply resulted in stiff fines and the possibility of a temporary shutoff of electricity service. The electricity shortage limited economic growth and had wide-ranging social consequences, including reduced energy services and public security. Diversifying energy supply would reduce the risk of power shortages or price shocks in the future.

Reduce Energy Sector Investments

Investments in the energy sector averaged about 9 percent of total capital investments in Brazil during the 1990s. Much of the investment in energy supply is now provided by the private sector or by profitable state-owned (or partially state-owned) companies such as the national petroleum company. But some of the investment still comes from the public sector. Energy supply is capital intensive and draws resources away from other critical areas including investments in health care, education, and housing. Reducing the total investment associated with meeting future energy service needs could benefit Brazil economically and socially.

Reduce Dependency on Energy Imports

Net energy imports increased from 27 million metric tons of oil equivalent in 1985 to 51 million metric tons in 2000 in spite of substantial growth in domestic oil production (MME 2000). Energy imports are mainly in the form of petroleum and petroleum derivatives, but electricity and natural gas imports are increasing as well. Energy imports absorb hard currency and harm Brazil's trade balance. Brazil ran an overall trade deficit of about $5 billion per year on average from 1997 to

1999, with net imports of petroleum and petroleum derivatives costing $4.4 billion per year on average (MME 2000). Reducing energy imports by increasing domestic energy supplies could also provide economic and social benefits.

Increase the Efficiency of Energy Use

Brazil has had some success with increasing the efficiency of electricity use, as described in Chapter 4. But many industries, businesses, and households still waste energy because of inefficient industrial processes, equipment, vehicles, and buildings. For example, motors used in Brazil are inefficient by international standards, as well as oversized and poorly operated in many cases (Geller et al. 1998). In the residential sector, using more efficient appliances could cut electricity use by nearly 30 percent (Almeida, Schaeffer, and la Rovere 2001). Cogeneration, an efficient technique for providing both electricity and useful thermal energy, accounted for only about 4 percent of electricity generation in Brazil as of 2000, far below the cost-effective cogeneration potential (Schaeffer and Szklo 2001). Increasing the efficiency of energy use would save consumers and businesses money and reduce the risk of new energy shortages.

Develop and Deploy Renewable Energy Sources

Brazil has plentiful renewable energy resources including wind, solar, and bioenergy resources (Winrock International 1999). The fraction of total energy supply provided by renewable energy sources, while still very high, is declining, due in part to increasing petroleum and natural gas production and use. Expanding renewable energy utilization could help to diversify energy supplies, stimulate new industries, create jobs, and contribute to economic and social development of poorer rural regions.

Reduce Adverse Environmental Impacts

Studies show that air pollution from burning fossil fuels, particularly fuel use for transport, is killing and injuring thousands of Brazilians each year (Azuaga 2000). Hydroelectric development floods forests and agricultural lands and can dislocate inhabitants. Nuclear power generates radioactive waste. In addition, carbon dioxide (CO_2) emissions are rising and are contributing to global warming. Cutting pollutant emissions associated with energy supply and use will improve air quality and public health and provide other environmental benefits.

Contribute to Social Development

Around 2.2 million households (about 5 percent of all households in Brazil) did not have electricity service as of 1999 (PNAD 1999). Some low-income households earning less than $150 per month still rely on wood as a major energy source (Oliveira et al. 1998). Increasing access to and use of modern energy sources by all households would reduce social and regional inequality, create job opportunities in underdeveloped areas, and avoid destruction of forests for fuel.

A wide range of barriers prevent achievement of these objectives. These barriers include limited availability and delivery infrastructure for some energy efficiency and renewable energy measures, high costs for some of the newer energy technologies, lack of awareness, lack of capital or convenient financing, and regulatory obstacles. In addition to these commonplace barriers, which were discussed generically in Chapter 2, Brazil has experienced many decades of economic instability and high inflation. These conditions strongly discouraged life-cycle analysis and longer-term investing, leading to a "culture" that tends to minimize first cost (Geller et al. 1998).

Policy Proposals

A variety of policy initiatives are needed to overcome these formidable barriers—both "carrots" and "sticks" that steer the private sector toward meeting Brazil's long-term energy service needs in ways that minimize waste and provide broad social and economic benefits. The following 12 national energy policies are proposed as part of a Clean Energy Scenario for Brazil. After presenting the policies, the Clean Energy Scenario is compared to a Base Scenario, which assumes a continuation of current policies and trends. The analysis examines energy supply and use as well as CO_2 emissions in Brazil through 2010, considering implementation of the 12 policies in combination.[6] While the analysis is not as extensive as that carried out for the United States, it indicates how the energy policies can contribute to the various objectives listed above.

Adopt Minimum Efficiency Standards for Appliances, Motors, and Lighting Products (Policy 1)

Minimum efficiency standards could be adopted for all new major household appliances (refrigerators, freezers, clothes washers, and air conditioners), lighting products (lamps and fluorescent lighting ballasts), and motors sold in Brazil after a certain date. Consumers would automatically purchase relatively efficient products without the need to educate and convince them to do so. By making more efficient products the norm, economies of scale occur and the incremental cost for greater efficiency is reduced.

The national electricity conservation program (PROCEL), together with the national testing and standards agency, already established energy efficiency test procedures and an efficiency labeling program. Also, PROCEL provides recognition and promotion of top-rated energy-efficient products. This will facilitate the adoption of minimum efficiency standards. In September 2001, the Brazilian Congress adopted legislation that authorizes and directs the federal government to establish mandatory minimum efficiency standards for appliances, motors, and lighting products. If this new law is swiftly acted upon, minimum efficiency standards could take effect relatively quickly.

It is reasonable to assume that Brazilian efficiency standards could be similar to those adopted in North America. Efficiency standards might provide 20 to 30 percent energy savings on average for new refrigerators, freezers, air conditioners, and lighting products (COPPE 1998, Geller et al. 1998). For motors, efficiency standards might yield electricity savings of 2 to 8 percent depending on the size of the motor (Tabosa 1999).

Expand Utility Investments in End-Use Energy Efficiency (Policy 2)

Starting in 1998, the federal regulatory agency for the electric sector (ANEEL) began requiring distribution utilities in Brazil to invest at least 1 percent of their revenues in energy efficiency programs. But only one quarter of the 1 percent must be spent on efforts that help consumers use electricity more efficiently. In 2000, the requirement was changed with half of the 1 percent devoted to a R&D fund, but the minimum amount for consumer-oriented efficiency programs was maintained (Jannuzzi 2001). Due to the power shortages in 2001, utilities spent about $80 million—approximately 0.5 percent of their revenues—on consumer-oriented efficiency programs in 2001 (Villaverde 2001).

This policy would expand funding for energy efficiency programs in Brazil to about 2 percent of utility revenues. It could be implemented through a revised mandate from ANEEL. Part of the money would be spent directly by the utilities and part would be directed to state and federal energy efficiency programs. The funding could be used to stimulate energy efficiency investments by households, businesses, and industries; provide financing for energy service companies; help establish the market for innovative energy efficiency measures; disseminate information; provide training, etc. Funding could be scaled up from current levels over a two-year period.

If this policy is adopted, it would be important to allow distribution utilities to recover the cost of these energy efficiency programs in utility rates. In addition, utilities should be given a financial incentive to operate effective programs that provide significant benefits for consumers and businesses. For example, utilities could be given a "bonus" equal to 10 to 20 percent of the net societal benefit generated by their efficiency programs, with an independent third party (e.g., ANEEL) evaluating this benefit utility by utility. The bonus could be collected through a small additional charge in utility bills that are paid by all consumers. Also, PROCEL could work with utilities to design effective programs and implement coordinated market transformation initiatives regionally or nationally (Geller 2000).

This policy is not duplicative of the minimum efficiency standards proposal. Certain technologies such as improvements in motor systems, better lighting designs, compact fluorescent lamps, alternatives to electric resistance water heaters, and air conditioning system improvements in commercial buildings cannot be implemented through standards since their feasibility depends on each application. Utility-funded energy efficiency programs could focus on promoting these systems-oriented measures.

Adopt Energy Codes for New Commercial Buildings (Policy 3)

No city or state in Brazil has adopted energy efficiency requirements for new commercial buildings. This policy would convene a group of experts to develop and publish a national model energy code that would include requirements for different climate zones in Brazil. The federal government could then give all municipalities over a certain size (say over 100,000 residents) a deadline for adopting the model energy code. This is a strong policy, but it is precisely what has been done in a number of Western nations.

Experience in other countries has shown that thorough training of architects, engineers, and builders is critical to the success of building energy codes, as is concerted monitoring and enforcement. So a key part of this policy would be to train builders, architects, building inspectors, and code enforcement officials from municipalities. PROCEL could carry out this effort, with energy experts from universities and technical institutes hired to conduct the training.

Commercial sector energy demand in Brazil grew nearly 8 percent per year on average between 1995 and 2000 (MME 2000). Commercial-sector electricity demand is projected to increase 6 percent per year in the future if energy codes and other policies to promote more efficient electricity use are not adopted (Eletrobrás 2000). It is assumed that this policy could eliminate 10 to 15 percent of the future growth in electricity demand in commercial buildings, including commercial buildings in the public as well as private sector (Lamberts 2001).

Expand Use of Combined Heat and Power Systems Fueled by Natural Gas (Policy 4)

There is substantial potential for using combined heat and power (CHP) systems in industries in Brazil. In addition, there is cogeneration potential in office buildings, hospitals, shopping centers, and other types of commercial buildings. The total cost-effective cogeneration potential is estimated to be in the range of 9 to 17 gigawatts (GW) of electric capacity (Eletrobrás 1999).[7]

However, the installed capacity of CHP systems in Brazil did not exceed 3 GW as of 2000. Most of this capacity was fueled by waste materials and waste gases in the paper, steel, and ethanol industries. Long-standing barriers that inhibit the development of CHP systems in Brazil include (1) relatively low electricity prices paid by larger industries; (2) lack of access to the utility grid for distributed, nonutility generators; (3) utilities unwilling to provide long-term power purchase contracts at reasonable rates; and (4) limited development and availability of natural gas resources (Soares, Szklo, and Tolmasquim 2001).

The recent increase in natural gas supply in Brazil opens up new opportunities for CHP systems. By the late 1990s, Brazil's natural gas supplies had increased appreciably, due mainly to the construction of the Bolivia–Brazil gas pipeline. In 1999, gas supplies reached about 32 million cubic meters per day. In addition, authorization was granted to import an additional 69 million cubic meters of natural gas from Argentina and Bolivia. Given the high efficiency of cogenerating electricity and useful thermal energy, the following policies could be adopted to eliminate the barriers to natural gas–based CHP systems.

1. Require utilities to purchase surplus power from CHP systems at avoided generation costs via long-term contracts, provided that these power supplies comply with certain reliability criteria.
2. Require utilities to interconnect CHP systems to the power grid without excessive delay or overly burdensome requirements, as well as provide back-up power to owners of CHP systems at reasonable terms.
3. Give priority to CHP projects as new gas supplies become available and are allocated to commercial and industrial consumers.

4. Provide financial incentives, such as long-term loans at attractive interest rates from the national development bank, for CHP systems that meet certain conditions such as high overall efficiency and low pollutant emissions.
5. Reduce import duties on CHP equipment such as gas turbines but also promote the production of this equipment in Brazil.

Approximately 10 percent of newly available gas supplies were allocated to CHP projects in 2001. This is a good start, but the other policies recommended here should be enacted as well. If this were done, it is reasonable to assume that at least 6 GW of new CHP capacity could be added by 2010 (Schaeffer and Szklo 2001). To analyze this policy, new CHP systems are assumed to have an electric-only efficiency of 35 percent, a total electric and useful thermal efficiency of 75 percent, and capacity factor of 70 percent on average.

Adopt Minimum Efficiency Standards for New Thermal Power Plants (Policy 5)

Brazil has sought for some time to increase electricity supplies from thermal power plants, but Brazil lacks high-quality coal reserves, and development of natural gas was quite limited until recently. The increased supply of natural gas has sparked great interest in the construction of natural gas–fired power plants. Many projects were proposed by utilities or private developers in recent years. Construction of gas-fired power plants has been slow, however, due to regulatory uncertainties and other factors. The slow pace of capacity expansion contributed to the electricity supply crisis in 2001, which in turn provided renewed impetus to remove these barriers.

The great majority of the gas-fired power plants proposed or under construction are simple cycle plants, meaning efficiencies of 30 to 35 percent rather than 50 to 60 percent achieved by state-of-the-art combined-cycle plants. Private investors prefer simple-cycle plants because of the lower investment costs, shorter construction time, and greater flexibility to respond to varying load conditions. In the future, some of these plants may be converted to combined-cycle operation.

Minimum efficiency standards could be adopted for all new gas-fired power plants that enter into operation in Brazil. Also, plants built as simple-cycle gas turbines could be required to add steam turbines and operate as combined-cycle plants if they are used more than a nominal amount. This policy would require all gas-fired power plants used over 500 hours per year to meet or exceed an efficiency level of 55 percent. This requirement would also narrow the difference in capital cost between electricity only and CHP plants, thereby helping to stimulate investment in CHP systems.

Adopt Industrial Energy Intensity Reduction Targets (Policy 6)

There is considerable potential to increase the efficiency of energy use in the industrial sector in Brazil by improving operating and management practices, using better equipment such as high-efficiency motors and motor-speed controls, and adopting innovative process technologies. One study indicates it is feasible to reduce energy use by 30 percent or more in a wide range of energy-intensive industries (COPPE 1998). This policy would establish energy intensity reduction targets for major industries in Brazil through voluntary agreements between the government and industry. Energy intensity targets would be analyzed and negotiated with specific industrial sectors.

To facilitate compliance and help companies meet the targets, PROCEL and the government could provide technical and financial assistance in the form of energy audits of industrial facilities, training, and tax incentives for investments in energy-efficient, state-of-the-art industrial equipment. Companies or sectors that enter into agreements to improve energy efficiency by at least 2 percent per year, and stay on track, could be protected from any increase in fuel taxes. Also, these companies or sectors could be given preferential access to power should electricity shortages recur. This policy is similar to the successful voluntary agreements effort in the Netherlands (see case study in Chapter 4) and also a proposal included in the policy recommendations for the United States.

It is assumed that this policy leads to a 12 percent reduction in overall industrial energy use by 2010. Some 80 percent of the savings might come, on average, from reducing fuel consumption and 20 percent from improving the efficiency of electricity use (Henriques and Schaeffer 1995).

Adopt Minimum Fuel Economy or CO_2 Emissions Standards for New Passenger Vehicles (Policy 7)

There are no fuel efficiency standards for new cars or light trucks in Brazil. Vehicle manufacturers receive some tax incentives for producing vehicles with engines 1 liter or smaller in volumetric capacity. Because of this policy, some 60 to 70 percent of all new passenger vehicles sold in Brazil have 1-liter engines. But the fuel efficiency of Brazilian cars and light trucks is still relatively low. In 1998, the average fuel economy of all passenger cars in circulation in Brazil was about

23.5 mpg or 10 kilomters per liter (km/l), while the average fuel economy of all new passenger cars sold that year in the country was about 26 mpg (11 km/l) (Azuaga 2000).

Passenger vehicles sold in Brazil are relatively inefficient because of the outdated technology employed in 1-liter Brazilian engines. Most of these engines are derived from 1.6-liter engines used to equip older models. But vehicle production by the multinational auto manufacturers is rapidly growing in Brazil. As production expands, it would be reasonable to insist that new vehicles include a variety of fuel-efficient technologies.

This policy calls for adopting passenger vehicle fuel efficiency standards in Brazil. These standards could be expressed in terms of either an increase in fuel economy (the approach followed in the United States) or a reduction in CO_2 emissions per kilometer traveled (as is the case in Europe). The advantage of a CO_2 emissions standard in Brazil is that auto manufacturers could opt either to raise fuel efficiency or produce and sell more ethanol (and other cleaner-fueled) vehicles. If a CO_2 emissions standard were adopted, manufacturers most likely would comply through some combination of efficiency improvement and fuel shifting.

The specific policy proposal is to require a 40 percent reduction in the average CO_2 emissions per kilometer for new passenger vehicles sold in Brazil by 2010, relative to the average emissions level in 2000. The standard would apply to the average emissions of domestic shipments by each manufacturer. It is assumed the standard is met by about a 5 percent per year reduction in average CO_2 emissions per kilometer starting in 2003. Furthermore, it is assumed that about 75 percent of this reduction will be achieved through efficiency improvement and 25 percent through increased sales of ethanol vehicles. By 2010, the policy leads to new vehicles with an average fuel economy of 16 km/l.

Expand the Production and Use of Ethanol Fuel (Policy 8)

Brazil's ethanol fuel program faces challenges, particularly as the fleet of neat ethanol vehicles produced during the 1980s is retired. The ethanol fuel mix as of 1999 was 54 percent hydrated ethanol, which is used in neat ethanol vehicles, and 46 percent anhydrous ethanol, which is blended with gasoline (MME 1999). The demand for ethanol will decline over the next decade unless policies to promote the purchase of new ethanol vehicles are strengthened and/or new outlets for ethanol and other sugarcane products are pursued.

This policy consists of a combination of actions to increase both the supply and demand for ethanol fuel over the next 10 years. First, low-interest loans could be offered to stimulate construction of new ethanol distilleries as well as expansion of existing distilleries. Second, the Brazilian government could create a "strategic ethanol reserve" of, say, 5 to 10 billion liters. The reserve would be tapped in case of shortfall between supply and demand. The national ethanol program suffered a setback in 1989–90 when demand exceeded supply and shortages occurred. Purchase of ethanol for the reserve could be paid for through a small tax on gasoline, or through a tax on both gasoline and ethanol fuel. For example, a gasoline tax of $0.005 per liter would provide enough revenue to purchase about 1 billion liters of ethanol per year for the reserve.

Third, new price or tax incentives could be provided to stimulate purchase of neat ethanol cars again. And fourth, ethanol could be blended with diesel fuel. Tests show that use of a 3 percent ethanol–97 percent diesel blend can be adopted without any engine problems, and in fact yields a substantial reduction in particulate emissions (Moreira 2000). The ethanol blend can be increased above this level (up to 12 percent) with use of a fuel additive to enhance the quality of the blend. This strategy has been adopted in Sweden (Moreira 2000).

In Brazil, a realistic set of targets in this area would include the following:

1. Ethanol accounts for 24 percent of the blend with gasoline.
2. A total of 6.5 billion liters of ethanol are purchased for the strategic reserve.
3. Sales of new neat ethanol cars start at 50,000 units and increase to 325,000 units by 2010.
4. The fuel economy of new ethanol cars starts at 10 km/l and rises to 13.3 km/l by 2010, consistent with the increase in fuel economy of gasoline cars.
5. Ethanol is blended with diesel fuel starting with 3 percent ethanol in the blend and increasing to 10 percent by 2010.

Stimulate CHP Systems Using Bagasse and Other Sugarcane Products (Policy 9)

Processing of sugarcane for ethanol or sugar produces a solid residue known as bagasse. Bagasse, which has significant energetic content, is burned to cogenerate electricity and steam in ethanol distilleries. But, as noted in Chapter 4, this is done at low pressure and efficiency at the present time to satisfy internal energy needs of the distillery only.

There is considerable potential to generate excess electricity using more efficient power generation technologies such as higher-pressure boilers, condensation and extraction cycle steam turbines, and gasification and combined-cycle technologies (Moreira 2000). Some of these measures are starting to be implemented with about 400 gigawatt-hours (GWh) supplied to the power grid from sugar mills as of 2001. Supplying electricity to the grid is particularly attractive since sugar mills typically operate during May through November, the dry season with respect to hydropower production in most of Brazil. Excess electricity produced in sugar distilleries can "firm up" additional hydropower available during the wet season in most years.

About 300 million tons of sugarcane were grown in Brazil as of 1999–2000. Assuming an increase in ethanol production from about 12 billion liters in 2000 to 16.5 billion liters in 2010, the sugarcane harvest would increase to about 350 million tons in 2010.[8] Given traditional methods of sugarcane harvesting and processing, this would result in 94 million tons of bagasse by 2010.

Traditional manual sugarcane harvesting in Brazil involves burning off the leaves and tops from the plants before cutting and harvesting. This reduces the amount of biomass available and causes significant air pollution at the regional level. The use of mechanized harvesting with no preburning could boost the amount of biomass appreciably, while at the same time reducing air pollution. For São Paulo state, it is estimated that mechanized harvesting could be implemented in over 70 percent of the planted area.

A combination of policies could facilitate higher-efficiency bagasse cogeneration as well as encourage use of leaves and tops for energy production, where appropriate. Some of these policies are similar to those needed to stimulate CHP with natural gas (see Policy 4 above):

1. Require utilities to purchase excess power from sugar mills at avoided generation, transmission, and distribution costs via long-term contracts.
2. Require utilities to interconnect CHP systems to the power grid without excessive delay or unreasonable technical requirements.
3. Continue to develop and demonstrate more efficient technologies such as bagasse gasification and combined-cycle power generation in sugar mills.
4. Provide long-term loans at attractive interest rates to sugar mills that adopt more efficient CHP technologies.
5. Finance and support the gradual adoption of mechanical harvesting systems.

Adopting these policies could result in 2,400 megawatts (MW) of bagasse CHP capacity by 2005 and 6,300 MW by 2010. The latter value is equivalent to nearly 10 percent of the power capacity in Brazil as of 2000. This capacity is assumed to consist of a mix of current steam turbines with low efficiency, higher-efficiency steam turbines, and innovative gasification and combined-cycle plants toward the latter part of the decade.

Mechanical harvesting of sugarcane would enable recovery and use of leaves and tops for energy production, but it would reduce the number of workers employed in sugarcane production. In the Clean Energy Scenario, it is assumed that 60 percent of sugarcane continues to be cut manually in 2005 and 30 percent in 2010 to limit the adverse impact on employment. Recovery of leaves and tops could be done where mechanical harvesting is already employed, or where air pollution from burning sugarcane fields is a serious problem and thus mechanical harvesting is desired for air quality reasons. As long as the shift to mechanical harvesting is gradual at the same time sugarcane and ethanol production are expanding, many of the displaced workers should be able to find new jobs within the industry.

Stimulate Grid-Connected Wind Power (Policy 10)

Wind power has long been used on a small scale to pump water in Brazil. The use of wind power to generate electricity began in 1992 on Fernando de Noronha, an island off the northeast coast of Brazil. Several wind power projects in the size range of 1 to 10 MW were installed between 1992 and 1999. But wind power did not become well-established in Brazil during the 1990s as it did in other countries due to a variety of barriers including lack of regulations defining utility interconnection terms and buyback rates. Nonetheless, one multinational company is manufacturing and marketing large-scale wind turbines in Brazil.

Brazil has substantial wind power potential both in coastal areas near population centers and in some interior regions. It is estimated that the state of Ceara alone has 25,000 MW of wind power potential (AWEA 2002). In February 2001, ANEEL established standard buyback rates for wind power and other types of renewable power generation. The rate for wind power was set at $48/MWh. However, this did not appear to be adequate to stimulate commercial wind power projects. As the power crisis unfolded in mid-2001, additional steps were taken to encourage wind power development.

In July 2001, the government unveiled a short-term Emergency Wind Power Program. The goal is to implement 1,050 MW of wind power by December 2003. The long-term power purchase price was increased to about $57/MWh for wind projects approved before the end of 2001, falling to about $52/MWh for projects approved by the end of 2002 (Wachsmann and Tolmasquim 2002). And in 2002, a law was enacted that requires distribution utilities to pay 80 percent of the average retail electricity price for wind power over a 15-year period (Moreira

2002). As a consequence of these policies, many new wind farms were proposed or under construction as of mid-2002. Wind power is on the cusp of becoming a significant electricity resource in Brazil.

These policies could be extended to achieve an orderly expansion of wind power until at least 2010. Assuming some large-scale wind projects are installed in the next two years and further technical advances and cost reductions occur, it may be possible to implement on the order of 7,000 MW of wind power capacity by 2010. For comparison, Germany implemented about 6,000 MW of wind power capacity and Spain about 3,100 MW as of 2000 through similar policies during the 1990s (BTM Consult 2001).

Stimulate Renewable Energy Use in Off-Grid Applications (Policy 11)

A program known as PRODEEM installed about 5,700 solar photovoltaic (PV) systems in off-grid areas primarily in northern and northeastern Brazil.[9] PRODEEM purchases PV systems in bulk and provides them at no cost to end users via state and local agencies. However, many of these systems are not well maintained and are not operating properly, due to technical problems and the fact that they are provided at no cost (Lima 2002).

It would make more sense to develop a private sector PV supply infrastructure in Brazil by supporting solar energy entrepreneurs as well as providing attractive microfinancing and subsidies to households that are not yet connected to the power grid.[10] This policy could include low-interest loans and technical support for rural PV dealers who market, install, and service PV systems. Subsidies could be reduced over time as PV technology improves and its costs drop. This type of integrated strategy that addresses both supply and demand has proven successful in solar PV programs in other countries such as India and Japan.

Of the 2.2 million households that did not have access to electricity as of 1999, many are impoverished and living far from the electricity grid. It is not cost-effective to extend the grid to these households given the high cost of rural electrification and low potential power demand of these households. This policy, if aggressively implemented, could lead to as many as half of these households obtaining solar PV systems by 2010. Furthermore, the focus could be on providing electricity for both domestic use (lighting, communications, entertainment, etc.) and productive purposes (home businesses and cottage industries) to foster social and economic development in poorer regions.

Improve the Efficiency of Freight Transport (Policy 12)

There are a number of technical options for increasing the efficiency of medium-duty and heavy-duty trucks, including more efficient engines, aerodynamic drag reduction, low-friction drivetrains, and reductions in energy expended for idling (Interlaboratory Working Group 2000). Likewise, there are various technical options for increasing the energy efficiency of trains. The policies that could stimulate these efficiency improvements in a country like Brazil include R&D and demonstration programs, tax incentives to encourage production and purchase of higher-efficiency trucks and locomotives, and, if necessary, fuel efficiency standards for new trucks. Considering that there will be some delays in introducing new technologies in Brazil, it is assumed that these policies yield fuel economy improvements of 16 percent for freight trucks and 12 percent for rail transport by 2010.

It is also possible to improve the energy efficiency of freight transport by shifting cargo among modes, specifically through shifting transport from less efficient trucks to more efficient trains and barges. In fact, the fraction of freight shipped by truck declined about 5 percent from 1996 to 2000 as rail and water transport services were expanded and improved (GEIPOT 2001). By continuing to invest in railroads, waterways, and intermodal freight transfer infrastructure, it might be possible to increase the fraction of freight shipped by rail from 21 percent in 2000 to around 29 percent in 2010. Likewise, the fraction shipped by water might rise from 14 percent to 18 percent. This means the fraction shipped by truck would fall by from 60 percent in 2000 to around 48 percent in 2010.

Energy and Other Impacts

The Base and Clean Energy Scenarios were analyzed using a computer model known as the Integrated Model for Energy Planning (IMEP).[11] IMEP provides integrated analysis of measures that affect both energy demand and supply. IMEP includes a high degree of end-use disaggregation and specificity, thereby enabling analysis of changes in the efficiency of appliances, vehicles, industrial processes, and the like. However, IMEP only analyzes energy supply and associated CO_2 emissions through 2010, and it does not include energy costs and other economic parameters other than GDP growth (Tolmasquim and Szklo 2000). Both the Base and the Clean Energy Scenarios were modeled assuming the same level of economic growth—4.7 percent per year on average from 2001 to 2010.

This is an optimistic growth rate that in reality will be difficult to achieve. However, this growth rate is considered to be possible in Brazil.

Table 6-1 shows the overall primary energy supply in 2000, 2005, and 2010 in both scenarios. In the Base Scenario, total energy supply increases 80 percent, or 6.0 percent per year on average from 2000 to 2010. Annual increases by energy type are: oil—3.1 percent, natural gas—20.3 percent, hydropower—1.8 percent, biomass—0 percent, and coal—2.6 percent. Natural gas use grows very rapidly due to its low starting point, introduction of gas imports, and rapid expansion of gas-fired thermal power plants during this time period. Total biomass use remains flat due to a slight reduction in charcoal and wood use offset by a slight increase in the use of sugarcane products.

In the Clean Energy Scenario, the policies that increase energy efficiency limit the growth in total primary energy use between 2000 and 2010 to 39 percent or 3.4 percent per year on average. Growth in oil use is limited to 1.0 percent per year, about one-third the rate in the Base Scenario. Growth in natural gas use is still rapid—about 16 percent per year on average—but total gas use in 2010 is about 32 percent less than in the Base Scenario. Total biomass use remains nearly flat in the Clean Energy Scenario, but use of sugarcane products increases more rapidly than in the Base Scenario, while the decline in wood and charcoal use is also greater.

Total renewable energy supply is about the same in the two scenarios. But given the differing rates of growth in total energy use, the share of primary

TABLE 6-1

Primary Energy Supply in the Base and Clean Energy Scenarios (million metric tons of oil equivalent)

	2000	2005		2010	
Energy Source		Base	CE	Base	CE
Petroleum and derivatives	87.9	100.2	94.4	119.7	97.2
Natural gas	9.8	44.1	30.8	62.2	42.5
Coal	13.4	15.6	14.0	17.4	14.0
Nuclear[a]	1.8	3.3	3.3	3.3	3.3
Subtotal—nonrenewable	(112.8)	(163.2)	(142.5)	(202.6)	(156.9)
Hydropower[a]	99.1	113.3	115.2	118.7	119.4
Wood and charcoal	21.4	20.9	19.8	20.4	17.7
Sugarcane products	22.2	23.4	24.9	23.5	24.9
Other	3.9	4.4	4.3	5.1	4.6
Subtotal—renewable	(146.6)	(162.0)	(164.2)	(167.6)	(166.7)
TOTAL	259.4	325.2	306.7	370.2	323.6

NOTES: [a]Nuclear power, hydropower, and other renewable electricity sources are counted in terms of equivalent fuel input for a central station thermal power plant.

energy provided by all renewable sources is 51 percent in 2010 in the Clean Energy Scenario compared to 45 percent in the Base Scenario. Thus the policies prevent the renewable energy fraction from falling as rapidly as expected under current trends. The reduction in renewable energy share in both scenarios is caused mainly by the growth in natural gas use.

Restraining growth in oil and gas consumption will have a positive impact on Brazil's trade balance. In the Clean Energy Scenario, projected oil production in 2010 exceeds internal demand for petroleum products by about 24 percent, thereby enabling Brazil to export crude oil or oil products. In the Base Scenario, projected oil production in 2010 approximately equals demand for oil products. In the case of natural gas, imports rise rapidly and account for 62 percent of total demand in 2010 in the Base Scenario. Imports account for 44 percent of total gas demand in 2010 in the Clean Energy Scenario. Thus gas imports in 2010 in the Clean Energy Scenario are much lower than in the Base Scenario.

Figure 6-3 shows the evolution of overall energy intensity (E/GDP) in each scenario. In the Base Scenario, energy intensity remains steady from 2000 to 2005 but falls about 10 percent by 2010. In the Clean Energy Scenario, energy intensity falls throughout the decade, dropping over 21 percent by 2010 due to increasing energy efficiency and CHP use. For comparison, overall energy intensity in Brazil increased nearly 15 percent between 1990 and 2000 due to residential energy services growing faster than economic output and other structural changes within the economy (Tolmasquim et al. 1998).

Figure 6-4 shows total electricity demand in 2000, 2005, and 2010 in each

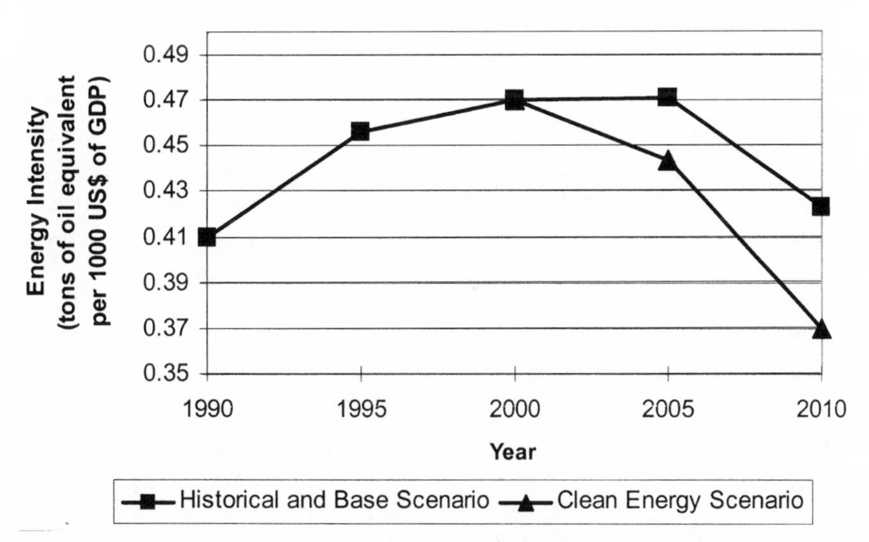

FIGURE 6-3: Energy intensity trends in the Base and Clean Energy Scenarios.

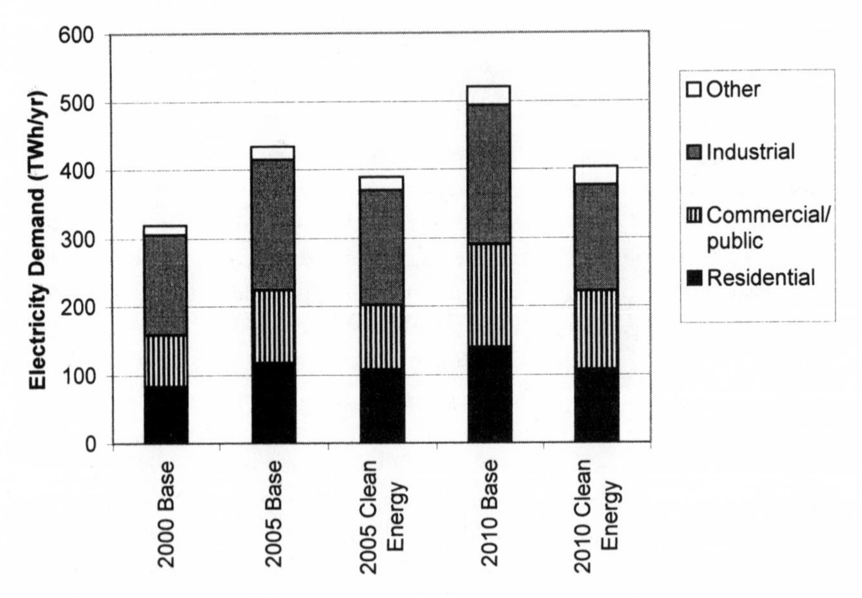

FIGURE 6-4: Electricity demand by sector in the Base and Clean Energy Scenarios.

scenario, broken down by sector. In the Base Scenario, electricity supply and demand increase about 5 percent per year on average from 2000 to 2010. Energy efficiency improvements in all sectors limit the growth to 2.5 percent per annum in the Clean Energy Scenario.

The electricity supply mix diverges in the two scenarios due to combination of lower demand growth and stimulation of CHP systems and renewable sources in the Clean Energy Scenario. CHP fueled both by natural gas and by sugarcane products accounts for 15 percent of total electricity supply in 2010 in the Clean Energy Scenario. For comparison, CHP provides only 6 percent of electricity supply in 2010 in the Base Scenario. Renewable sources, including hydropower, wind and solar power, and biomass cogeneration provide 67 percent of electricity generated in 2010 in the Clean Energy Scenario compared to 56 percent in the Base Scenario.

A total of 39 GW of new central-station gas plants are added between 2000 and 2010 in the Base Scenario compared to about 26 GW in the Clean Energy Scenario. In contrast, about 12 GW of CHP capacity is added in the Clean Energy Scenario compared to about 4 GW in the Base Scenario. Combining these two resources, there is about 10 percent less overall expansion in power-generating capacity in the Clean Energy Scenario. Also, there will be less need for

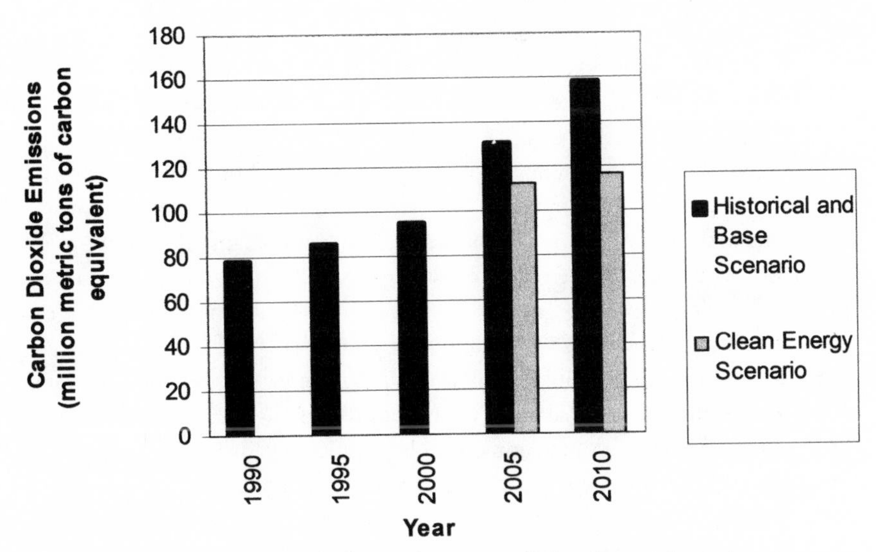

FIGURE 6-5: Carbon dioxide emissions in the Base and Clean Energy Scenarios.

transmission and distribution system investment because new capacity will be more decentralized and closer to power demand. Thus the required investment in the power sector should be less in the Clean Energy Scenario.

Figure 6-5 shows the evolution of CO_2 emissions from the energy sector in Brazil for each scenario. In the Base Scenario, CO_2 emissions grow by 66 percent from 2000 to 2010.[12] In the Clean Energy Scenario, the growth in CO_2 emissions is limited to 23 percent over this decade. Consequently, implementing these policies would contribute to the global effort to limit greenhouse gas emissions and global warming. Implementing the policies could also provide opportunities for project cofunding through the Clean Development Mechanism of the Kyoto Protocol (e.g., for the incremental costs associated with installing wind power, solar PV, or bagasse cogeneration systems). The Clean Development Mechanism is discussed in Chapter 7.

Summary

The policies proposed in the Clean Energy Scenario would provide a broad range of benefits and contribute to meeting nearly all of the objectives discussed above. First, the policies should reduce investment requirements in the energy sector due to efficiency improvements and consequently lower energy demand growth, although this issue needs further analysis. Some investments would be shifted from energy producers to end users (consumers and businesses) who would purchase more efficient products, vehicles, and CHP equipment. If these investments are made at the time of equipment replacement, the incremental cost should be moderate (Geller et al. 1998).

The policies clearly will reduce net energy imports and thus improve Brazil's trade balance. Natural gas imports will be reduced due to efficiency improvements and reduced need for natural gas–fueled power plants. In addition, oil production could exceed demand for oil products if domestic oil production grows as anticipated. These impacts could be quite significant, providing a net energy trade surplus of around $5 billion per year by 2010.

By design the policies would increase energy efficiency and expand renewable energy use in Brazil. National energy use would be about 12.5 percent lower by 2010 if the policies are adopted and effectively implemented. Total renewable energy use in 2010 would be about the same with or without the policies, but the policies would result in greater supply of ethanol, wind power, solar power, and electricity produced from sugarcane products. The increase in these modern renewable energy sources would be offset by reduced use of traditional energy sources (wood and charcoal) in the Clean Energy Scenario. Also, the policies result in renewable sources providing a higher fraction of total energy consumption than in the Base Scenario.

The policies clearly would have a number of positive environmental impacts, although only the impacts on CO_2 emissions were quantified. The policies would greatly reduce the growth in CO_2 emissions compared to what is anticipated given current trends. The policies would reduce particulates and NO_x emissions from vehicles due to the lower reliance on diesel-fueled trucks and higher efficiencies of all types of vehicles, thereby improving urban air quality and public health. The policies would also result in less air pollution from uncontrolled burning of sugarcane leaves and tops.

The policies proposed in the Clean Energy Scenario would have positive social impacts as well. One of the policies is designed to expand electricity provision to impoverished rural households not currently using electricity. The policies would greatly expand both solar PV and wind power applications in the northern and northeastern regions of Brazil, areas currently exhibiting high levels of poverty and social underdevelopment. And the policies would expand and increase the competitiveness of the sugarcane and ethanol industries, which are very labor intensive compared to other industries in Brazil.

The policies provide some degree of energy supply diversification through expansion of wind power and increased use of sugarcane products. But the policies also reduce growth in natural gas and coal use compared to what is expected if current trends are maintained. With less overall energy use in the Clean Energy

Scenario, petroleum and hydropower combined provide 67 percent of total energy consumption in 2010 compared to 64 percent that year in the Base Scenario and 73 percent in 2000. Thus the policies would not result in greater overall energy supply diversification relative to what is expected given current trends.

Brazil has demonstrated the ability to adopt and effectively implement innovative energy policies and technologies as exemplified by the ethanol fuel program and efforts to increase the efficiency of electricity use. These efforts involved a long-term commitment from the government; a comprehensive set of policies to overcome technical, institutional, and market barriers; and active engagement of the private sector. A similar strategy could be used to successfully implement the policies proposed here.

A few of the proposed polices include new financial incentives, especially those aimed at deploying renewable energy sources. These incentives could be paid for through higher taxes on conventional fossil fuels such as petroleum products. In fact a higher tax on gasoline has been suggested in Brazil to fund new incentives for expanding ethanol production and use. Other policies include market obligations such as paying avoided costs for electricity from CHP systems and offering attractive feed-in payments for electricity produced by wind power projects. If these policies create significant distortions in the emerging competitive electricity supply market in Brazil, their incremental costs could be shared among all electricity suppliers or consumers.

The policies proposed for Brazil could also have application and value in other developing countries. These countries have many of the same needs as Brazil—providing expanded energy services at least cost, increasing the efficiency of energy supply and demand, reducing fossil fuel imports and making greater use of domestic renewable energy resources, and improving the living conditions and economic opportunities of impoverished citizens. The magnitude and priority of these needs will vary from country to country. Nevertheless, many developing nations could benefit from implementing similar policies, tailored of course to their own resources, capabilities, and needs.

Notes

1. This chapter presents monetary units in U.S. dollars based on an exchange rate of 2.35 Reais per U.S. dollar, the level in January 2002.
2. This figure includes only half the capacity of the 12.6 GW Itaipu binational hydro plant even though Brazil consumes nearly all of the output from this plant (half the ownership belongs to Paraguay).
3. These figures value hydroelectricity based on the fuel input needed to

produce an equivalent amount of power in Brazilian thermal power plants, which have an average efficiency of 27.5 percent. This is the convention used in Brazil.

4. The energy sector includes oil refineries, ethanol distilleries, and power plants.
5. This is my own set of energy policy objectives, but they reflect broader national goals in Brazil, including maintaining strong economic growth, fostering social development in impoverished communities and regions, and reducing environmental degradation.
6. Unfortunately, I was not able to analyze the economic costs and benefits of the 12 policies.
7. For comparison, Brazil had about 68 GW of total electric capacity as of 2000.
8. These values also assume there will be an increase in distillery productivity (ethanol output per ton of sugarcane) during the next decade.
9. PRODEEM is the Program for the Development of Energy in States and Municipalities. It is coordinated by the Brazilian Ministry of Mines and Energy.
10. This policy emphasizes solar PV systems and markets, but it could support other off-grid renewable energy technologies such as small-scale wind and bioenergy systems where appropriate.
11. The IMEP was developed by researchers at the Federal University of Rio de Janeiro (Tolmasquim and Szklo 2000).
12. Carbon dioxide emissions in 2000 are relatively low due to the large fraction of energy supply provided by renewable sources.

International Policies and Institutions

The preceding chapters discussed individual policies or groups of policies adopted at the national, state, and local levels. They show that national and state policies are the primary mechanisms for advancing energy efficiency and renewable energy use worldwide. But there is also a role for international policy and policy coordination, as well as for support from international agencies such as the International Energy Agency, the World Bank, and the United Nations. International policies that affect clean energy development worldwide include:

- cooperation on technology research, development, and demonstration (RD&D), standards, and deployment among countries;
- policies that influence technology transfer to developing countries;
- the energy programs of bilateral development assistance agencies, the Global Environmental Facility, and the United Nations;
- the energy policies and lending portfolios of the World Bank and other multilateral development banks; and
- the global climate change treaty.

Each of these topics is addressed in the following text. This chapter concludes by asking if current international institutions are adequate for supporting a transition to a more sustainable energy future, and proposes a new agency that could hasten that transition.

International Clean Energy Cooperation

Cooperation among countries is needed to facilitate the transition to a cleaner, energy-efficient future. International cooperation in RD&D on new energy technologies can leverage resources and increase the pace of technological innovation. Cooperation with respect to policies such as test procedures, labeling, efficiency standards, and financial incentives can minimize differences among countries and facilitate responses by the private sector. Harmonizing policies is especially important for mass-produced products that are manufactured and sold on an international scale (e.g., auto manufacturers could benefit from harmonized fuel economy test procedures and standards, as could worldwide energy efficiency efforts, assuming reasonable standards were adopted).

Outside of activities strictly in the private sector, worldwide energy technology cooperation is orchestrated mainly by the International Energy Agency (IEA). The IEA was established in the wake of the 1973 oil crisis and has 26 member countries from Europe, North America, and the Pacific region. The IEA's primary mission is to promote the effective operation of international energy markets as well as maintain a system for coping with oil supply disruptions (IEA 2001d).

The IEA also conducts or coordinates a wide range of activities to support greater energy efficiency and renewable energy use worldwide. The IEA analyzes and recommends policies to promote energy efficiency and renewable energy use in member countries, advises on energy technology deployment, and encourages international cooperation in RD&D and dissemination of innovative energy technologies. In addition, the IEA promotes more environmentally sustainable provision and use of energy (IEA 2001d).

Much of the IEA's work is done through implementing agreements that individual member countries voluntarily participate in and contribute to. The parties to these agreements jointly undertake energy technology RD&D, demonstration projects, technology or market assessments, procurement initiatives, and information exchange and dissemination. Examples of these agreements include:

- collaborative RD&D on renewable energy technologies, heat pumps, and hydrogen production and use;

- collaborative design of improved district heating and cooling systems;
- coordinated procurement of innovative energy-efficient appliances, lighting products, office equipment, and other technologies;
- information exchange on technologies and programs for demand-side management and renewable energy implementation; and
- development of test procedures for evaluating the performance of wind turbines and other innovative energy technologies.

The IEA also hosts the Climate Technology Initiative, which is assisting developing countries with the adoption of environmentally sound energy technologies and practices. The Climate Technology Initiative supports capacity building, information dissemination, and technical assistance activities (IEA 2001h). However, the budget of the Climate Technology Initiative is quite small.

Energy technology and policy cooperation is increasing within the European Union. The voluntary agreement between the European Commission and vehicle manufacturers to improve the efficiency and reduce the carbon dioxide (CO_2) emissions of automobiles, and European appliance labeling and standards initiatives, were described in Chapter 3. In addition, the European Commission and manufacturers entered into a voluntary agreement to reduce the electricity use of electronic products such as televisions and VCRs. The IEA supported this initiative by organizing workshops and raising awareness concerning standby power. This effort contributed to the 50 percent decline in the standby power consumption of televisions sold in Europe between 1995 and 1999 (IEA 2001f).

In 2000, the European Commission adopted a directive intended to double the contribution of renewable energy as a fraction of primary supply in the European Union by 2010 (European Commission 2000). The directive includes indicative targets for renewable electricity supply in each country (Table 7-1). Taken together, these targets would lead to renewable sources providing 22 percent of all electricity consumed in the European community by 2010. The directive requires each country to set national renewable energy targets and adopt national policies for meeting these targets. The directive also establishes a renewable electricity certification system, and it allows countries to continue their national renewable energy subsidy schemes for at least 10 years so as not to hinder deployment.

Regional energy cooperation is on the rise in other parts of the world as well. In some cases this cooperation involves transferring technologies and strategies that one country has successfully implemented to nearby countries. For example, the United States has helped Mexico develop appliance efficiency standards and commercial building energy codes (Huang et al. 1998). Sweden is helping countries in the Baltic Sea region increase the energy efficiency of their district heating systems and convert these systems to bioenergy sources. This project involves both

TABLE 7-1

Indicative Targets for Electricity Consumption from Renewable Sources in the European Union

Country	2010 Renewable Electricity Target (TWh)	1997 Renewable Electricity Fraction (%)	2010 Renewable Electricity Fraction (%)
Austria	55.3	72.7[a]	78.1
Belgium	6.3	1.1	6.0
Denmark	12.9	8.7	29.0
Finland	33.7	24.7	35.0
France	112.9	15.0	21.0
Germany	76.4	4.5	12.5
Greece	14.5	8.6	20.1
Ireland	4.5	3.6	13.2
Italy	89.6	16.0	25.0
Luxembourg	0.5	2.1	5.7
Netherlands	15.9	3.5	12.0
Portugal	28.3	38.5[a]	45.6
Spain	76.6	19.9	29.4
Sweden	97.5	49.1[a]	60.0
United Kingdom	50.0	1.7	10.0
European Union	675.0	13.9	22.1

NOTES: [a]Austria, Portugal, and Sweden obtain over 30 percent of their electricity from hydropower.
SOURCE: European Commission 2000.

technical assistance and subsidized financing. Over 70 projects were completed or under way as of 2001. The program has led to the commercialization of biomass-fueled boilers in the Baltic region as well as to the improvement of local skills in energy efficiency and renewable energy technologies and deployment (IEA 2001g).

Cooperative efforts are emerging within the developing world as well. For example, the 14 countries of the Southern African Development Community have initiated a program in clean energy technology cooperation. This initiative is focused on technology assessment, building human and organizational capacity, and attracting investment in clean energy technologies within the region. Likewise, the African Energy Policy Research Network (AFREPREN) has brought together academics and policy makers to develop innovative energy strategies (Karekezi 2002b).

International clean energy technology cooperation faces a number of challenges. Some countries are focused on promoting their own industries and exports, rather than supporting manufacturing and distribution of clean energy technologies in other nations. Also, it can be difficult to reach agreement among countries on common policies due to competitive pressures. For example, European nations have had a difficult time adopting harmonized appliance efficiency labels and standards due to different national perspectives and interests (IEA 2000b). But the potential benefits from international collaboration suggest that nations should expand multinational efforts to develop and deploy energy efficiency and renewable energy technologies (Box 7-1).

BOX 7-1

U.S. Support for International Energy Cooperation

The U.S. federal government spent about $235 million on 320 different projects that could be considered international energy technology collaboration as of 1997 (PCAST 1999). These include projects pertaining to fossil fuel technologies and nuclear power as well as energy efficiency and renewable energy technologies. In fact, collaboration on nuclear fission and fusion accounted for 60 percent of U.S. support for international energy cooperation as of 1997. Collaboration on energy efficiency and renewable energy accounted for less than 20 percent of the total.

The U.S. government and other U.S. institutions contributed to a number of the clean energy "success stories" in spite of these skewed priorities. For example, the U.S. government and some charitable foundations provided start-up funding for the energy efficiency centers in Russia and Eastern Europe discussed in Chapter 3 (Chandler et al. 1999, IEA 2001g). The U.S. government also provided valuable support to the PROCEL energy efficiency program in Brazil and to successful clean energy efficiency initiatives in other countries such as China, India, the Philippines, and Thailand (Geller 2000, IEA 2001g).

Because of the importance of global energy use to the U.S. economy, national security, and the environment, a panel of advisers to President Clinton recommended increasing U.S. funding for international energy technology collaboration by a factor of 3 (an increase of $500 million per year) within five years (PCAST 1999). The panel suggested that the bulk of the additional funding be devoted to international collaboration in energy efficiency and renewable energy technologies, and that the United States increase its support for capacity building, energy technology innovation and energy sector reform worldwide, and reform of energy lending by the international financial institutions. The panel also recommended establishing a Strategic Energy Cooperation Fund to coordinate these efforts.

◉ ◉ ◉

Fostering Clean Energy Innovation in Developing Countries

Developing countries have some features that could enable them to be leaders in the transition to a clean energy future. These features include plentiful renewable energy resources and energy efficiency opportunities; nascent industrial, transport, buildings, and power infrastructures; and high growth rates in energy production and energy-intensive activities (Reddy, Williams, and Johansson 1997). As developing countries progress economically and socially, there is the potential to "leapfrog" over the inefficient, fossil fuel–based, and polluting energy production and consumption patterns found in industrialized nations (Goldemberg 1998).

The Brazilian ethanol fuel program described in Chapter 4 provides one example of leapfrogging. It was successful because it built on the well-established sugar industry in Brazil, it relied on the private sector for fuel production, it fostered steady technological and agricultural advances, and it received strong governmental support. There are many other leapfrogging opportunities in developing countries, including use of distributed renewable power technologies for rural electrification, use of biogas plants to produce cooking fuel, adoption of compressed natural gas, electric and fuel cell vehicles as motorized vehicles are introduced, and use of high-efficiency, low-polluting manufacturing processes as industries expand. For example, developing countries could adopt cleaner, state-of-the-art technologies for steel production, such as high-efficiency electric arc furnaces, thin slab casting, and smelt reduction as their steel industries expand. In fact, South Korea and South Africa installed the first smelt reduction steel plants that are operating in the world[1] (Phylipsen et al. 1999).

The policies that can foster energy technology innovation and leadership in developing countries include: (1) RD&D that emphasizes clean energy supply and energy end-use technology innovation in developing countries, (2) development of new industries and introduction of new technologies through international joint ventures and other technology transfer mechanisms, (3) adopting and enforcing strong energy efficiency and environmental standards so that new infrastructure is state-of-the-art rather than technologically outdated, and (4) attractive financing and market development assistance to clean energy technology entrepreneurs, including small- and micro-scale enterprises (Goldemberg 1998, Karekezi 2002b).

Technology transfer between industrialized and developing countries can be an important component of clean energy development worldwide. In considering technology transfer mechanisms, it is important to recognize that investment by private companies represents a large and growing share of total financial flows to developing countries. Official development aid represented only about 15 percent of net financial flows to developing countries in 1997 compared to about 43 percent in 1990. Of the roughly $48 billion in development aid in 1997, technical cooperation represented about $13 billion. In contrast, investment in developing countries by private companies reached about $110 billion in 1997 (Goldemberg 1999).

Private sector investment in the energy sector of developing and former communist nations has supported modernization and reduced overall energy intensity to some degree. China, for example, opened its power sector to foreign investment in the early 1990s. The 20 or so new power plants that were either operational or under construction as of 1998 were significantly more energy efficient than typical power plants operating in China (Blackman and Wu 1999). About one-fifth of these plants featured cogeneration of power and thermal energy. Looking across developing countries, there has been a strong correlation between increasing foreign investment and declining energy intensity over the past 15 years (Mielnik and Goldemberg 2002).

Given the growing importance of private sector investment in developing countries, encouraging joint ventures and licensing can be effective strategies for disseminating clean energy technologies. For example, Danish wind turbine manufacturers have established successful joint ventures in India. Shell Solar and ESKOM (a state-owned utility in South Africa) have established a joint venture to assemble, market, and service off-grid photovoltaic (PV) systems in South Africa. In the case of smelt reduction steel mills built in South Korea and South Africa, the technologies were purchased from specialized suppliers based in industrialized countries (de Beer, Worrell, and Blok 1998). Protecting intellectual property rights can facilitate the flow of innovative energy technologies to developing countries (Goldemberg 1998, PCAST 1999).

Joint ventures and licensing can help developing countries rapidly obtain state-of-the-art technologies. For some technologies, manufacturing, assembly, and marketing can be carried out locally by joint venture partners or licensees. A substantial number of local jobs can be created through local manufacturing or assembly, which in turn can help to reduce costs. Although it may be necessary to import key components at first, nationalization should occur as markets grow. In the joint venture between Shell Solar and ESKOM, for example, the solar panels will be manufactured by Shell in Europe, but other system components are being made and assembled in South Africa. Governments can encourage joint ventures and licensing of sustainable energy technologies and services by providing financing, tax incentives, and market development support.

As energy efficiency and renewable energy technologies become established in developing countries, it is important to strike a balance between maximizing sales and addressing the needs of poorer segments of a society. Markets can grow more quickly by serving businesses and wealthier households, as demonstrated in solar PV projects in India and elsewhere (Miller and Hope 2000). While it is reasonable to serve these markets to build up local manufacturing and/or service capacity, it should not be the exclusive focus. Providing efficient, modern energy services to poorer households, including those in rural areas, can provide enormous social benefits (Box 7-2). The strategies for serving low-income, rural communities include targeted infrastructure development, providing microfinancing and subsidies to rural households, and involving local communities, in particular local women, in program design and implementation (Goldemberg 2000, Martinot et al. 2002).

BOX 7-2

Energy and Social Issues in Developing Countries

There are strong links between energy production and use on the one hand and social conditions on the other. Put simply, lack of modern energy sources contributes to poverty, poor health, undereducation, unemployment, and high population growth. Conversely, provision of clean, efficient, and modern energy sources especially in rural areas can

- reduce burdensome labor requirements for collecting firewood and water, and reliance on human labor for farming;
- enhance educational opportunities by diminishing the need for child labor and providing electricity for lighting, telecommunications, and other services;
- improve public health, especially the health of women and children, by cutting the need for "backbreaking" human labor, improving indoor and outdoor air quality, and improving water quality and sanitation;
- create job opportunities in energy efficiency and renewable energy industries, as well as other areas; and
- lower population growth by improving education levels and reducing the need for child labor.

Thus, providing renewable-based fuels and electricity, improved cookstoves, mechanical water pumping, and mechanized farming equipment should be given high priority in national and regional economic and social planning, not just energy planning. Likewise, energy policies and programs should pay special attention to serving the needs of low-income households in both urban and rural areas. This means ensuring that low-income households have access to renewable energy devices and efficient lighting, appliances, dwellings, and transport services (Karekezi 2002b).

The needs and capabilities of women are often overlooked in efforts to improve energy efficiency, disseminate renewable energy technologies, and promote more sustainable transport systems. Women are generally responsible for cooking and obtaining traditional cooking fuels in developing countries, as well as other household and agricultural tasks (Misana and Karlsson 2001). Increasing the participation of women in energy policy making as well as energy program design and implementation can enhance sustainable energy development, poverty alleviation, and gender equity (Cecelski 1995, Shailaja 2000).

Focusing on women in rural energy development can also enhance income-generating opportunities through rural energy enterprises, providing electricity or mechanical power for cottage industries, or improving the efficiency and output of food processing. For example, women have played a leading role in successful improved cookstoves, bioenergy development, and fuel-efficient food processing initiatives in various African countries (Misana and Karlsson 2001). In addition, women can play a leading role in the production and sale of energy efficiency and renewable energy technologies in developing countries.

◉ ◉ ◉

Bilateral Assistance, the Global Environmental Facility, and the United Nations

Most industrialized countries support clean energy development and deployment in Third World countries through their bilateral assistance programs. There are many examples of valuable bilateral assistance (see Box 7-2 for examples involving U.S. assistance). Other examples include the support from Sweden for bioenergy development and cogeneration systems in the Baltic nations, support from Germany for PV system commercialization in Kenya, and support from Denmark for wind power development in India and other developing countries (GEF 2001a, IEA 2001g).

Bilateral energy assistance is not always helpful, however. Many projects are politically motivated, are planned and managed by foreign specialists, or simply install and demonstrate technologies from donor nations (PCAST 1999). The distinction between development assistance and export promotion is frequently blurred. This approach can undermine rather than foster local capacity building and the creation of viable markets for clean technologies in developing countries (GEF 2001a).

It is especially important to nurture local businesses as part of donor-supported energy efficiency and renewable energy projects. In Zimbabwe, for example, a $7 million solar PV assistance program surpassed its goal of installing 10,000 solar PV systems between 1993 and 1997. But the project was not successful in building a self-sustaining solar PV supply network or market in the country. In fact, the project created market distortions that hurt some portions of the nascent local renewable energy industry operating in Zimbabwe prior to the start of the project (Mulugetta, Nhete and Jackson 2000). Unfortunately, many energy assistance projects have emphasized meeting installation targets rather than building conditions for replication and market expansion (Martinot et al. 2002).

Both donor agencies and local governments should learn from these experiences. Donor-supported programs need to be carefully designed and driven by local technological, socioeconomic, and institutional needs. Donor support should contribute to a long-term strategy to build sustainable markets for clean energy technologies in developing countries (Martinot et al. 2002). Developing countries should insist on these conditions when negotiating projects with potential funders.

The Global Environmental Facility (GEF) was created with funding from industrialized nations to help developing countries implement global environmental treaties. In the climate change area, the GEF funds projects to remove barriers to greater energy efficiency and renewable energy use, to reduce the long-term costs of cleaner energy technologies, and to promote more sustainable transport systems. GEF projects are implemented by the World Bank, the United Nations Development Programme, and the United Nations Environment Programme. The GEF allocated over $1 billion to around 270 specific climate change–related projects from 1991 to 1999 (GEF 2001b).

Some GEF projects help to leverage larger-scale loans from the World Bank for energy efficiency and renewable energy development (see Table 7-2). Also, the GEF is collaborating with the International Finance Corporation to support renewable energy businesses and market development in various developing countries (GEF 2001a). The GEF has contributed to a number of successful clean energy initiatives (Birner and Martinot 2002, GEF 2001b, Martinot and McDoom 2000). These include:

TABLE 7-2

Partial List of Global Environmental Facility (GEF) and World Bank Energy Efficiency and Renewable Energy Projects

Country	Project Type	GEF Funding (million $)	World Bank Funding (million $)
Argentina	Renewable energy technology	10	30
Brazil	Energy efficiency	15	43
Cape Verde	Renewable energy technology	5	18
China	Solar home systems	35	100
China	Energy service companies	22	63
India	Renewable energy technology	26	190
India	Energy efficiency	5	170
Indonesia	Solar home systems	24	20
Mauritius	Bioenergy	3	15
Philippines	Geothermal	30	227
Sri Lanka	Renewable energy technology	6	24

SOURCES: Martinot 2001, Martinot and McDoom 2000.

- renewable energy financing and market development through IREDA in India;
- energy efficiency programs and infrastructure development in China, Hungary, and Thailand;
- renewable energy programs and infrastructure development in China, India, Vietnam, Central America, and parts of Africa;
- energy-efficient lighting technology introduction and market development in Mexico, Thailand, and Poland;
- adoption of energy efficiency codes and standards in China, Thailand, and West Africa; and
- utility sector reforms that are enabling independent renewable energy developers to sell power to the electric grid in Mauritius and Sri Lanka.

The GEF is moving away from funding discrete technology-oriented projects and is increasing support for multifaceted efforts aimed at removing market barriers and establishing self-sustaining energy efficiency and renewable energy markets in developing countries (Birner and Martinot 2002). To further increase its effectiveness, the GEF could provide more resources for market transformation including policy reform and capacity building in both the private and the public sectors, and consequently less funding for implementation of specific technologies (IPCC 2000). In doing so, GEF and its funders will need patience since capacity and institution building may not provide immediate results in terms of renewable energy or energy efficiency devices installed. In addition, developing countries must be willing to support this market-oriented approach for it to be successful.

A number of branches of the United Nations support sustainable energy initiatives in developing countries. The United Nations Development Programme (UNDP) promotes innovative energy policies and funds capacity building, training activities, and feasibility studies. Its overall objective is to foster human development through greater energy efficiency, renewable energy use, and introduction of other modern, clean energy technologies (UNDP 2001). The United Nations Environment Programme (UNEP) hosts a Collaborating Centre on Energy and Environment. The Centre focuses on helping developing countries integrate environmental considerations into energy planning and policy making. It does this by conducting studies and supporting research by local institutions, coordinating projects, and disseminating information (UCCEE 2001). Both the UNDP and UNEP also develop and implement projects for the Global Environmental Facility. In addition, the United Nations Industrial Development Organization (UNIDO) supports industrial energy efficiency improvement and the United Nations Food and Agricultural Organization (FAO) addresses rural energy issues in developing countries as part of its mandate.

Multilateral Development Banks

The World Bank and regional development banks (known collectively as the multilateral development banks, or MDBs) are important lenders for energy projects in developing and transition countries.[2] Also, recipient nations are able to obtain further loans at attractive terms once the World Bank or regional development banks give their "stamp of approval" to a project. Historically, the vast majority of energy sector loans by the MDBs went to large-scale hydropower, fossil fuel, and other energy infrastructure projects. Very little financing went to energy efficiency or smaller-scale renewable energy projects. For example, only one-third of 1 percent of World Bank power sector lending between 1992 and 1996 was dedicated to increasing the efficiency of electricity use, in spite of the enormous potential for saving electricity at lower cost than supplying electricity in developing countries (Strickland and Sturm 1998).

Recently the development banks have begun to improve on this record. The World Bank indicated that between 1994 and 1998 it approved $1.2 billion in loans for end-use energy efficiency projects, efficiency improvements in district heating systems, and nontraditional renewable energy projects—equivalent to about 7 percent of total World Bank energy lending during this five-year period (World Bank 1998). The World Bank approved 17 projects with renewable energy components totaling about $700 million between 1992 and 1999, and the Global Environmental Facility provided $230 million in grants for these projects (see Table 7-2).

This is a positive trend, but the World Bank and other MDBs could do much more to facilitate a clean energy revolution in developing countries. As the markets for and capability to implement sustainable energy technologies in developing nations grow, the MDBs could phase out their funding for conventional energy projects and allocate all of these resources to energy efficiency, renewable energy, and natural gas projects. If developing countries prefer conventional energy technologies, project developers could obtain financing from commercial banks or other sources of capital.

To facilitate this "lending transition," the MDBs need to increase their own capacity and expertise in energy efficiency and renewable energy technologies (Martinot 2001). The Asia Alternative Energy Program (ASTAE) within the World Bank is a good model for this. ASTAE was created to foster energy efficiency and renewable energy projects within World Bank energy lending to Asia. With assistance from ASTAE, 18 energy efficiency or renewable energy loans were provided to 11 Asian nations between 1993 and 2000, in many cases together with grants from the Global Environmental Facility or bilateral donors for training, capacity building, and market development (World Bank 2000). These projects are expected to displace 1.5 GW of fossil fuel-based generation through a total investment of $3.8 billion in efficiency and renewable energy technologies.

The Climate Treaty

The United Nations Framework Convention on Climate Change was adopted by over 150 nations at the Rio de Janeiro "Earth Summit" in 1992. It was ratified by the United States and other countries, and entered into force in 1994. The ultimate objective of the Convention is to stabilize the concentration of greenhouse gases in the atmosphere at a level that would prevent dangerous interference with the world's climate. The Convention established the principles that industrialized countries should take the lead in reducing emissions because of their high historical emissions, and that industrialized countries should provide funding for technology transfer to poorer nations for both mitigation and adaptation efforts.

The Climate Change Convention included a nonbinding provision that industrialized countries attempt to return their greenhouse gas emissions to 1990 levels by the end of 2000. This provision led to many initiatives by industrialized nations to increase energy efficiency, renewable energy use, and a shift to natural gas during the 1990s (see examples for Denmark, Germany, Japan, the Netherlands, Spain, the United Kingdom, and the United States discussed in Chapters 3 and 4). However, most Western countries did not come close to meeting this target (Table 7-3).

TABLE 7-3

Greenhouse Gas Emissions of Industrialized Countries and Kyoto Protocol Targets

Country	1990 GHG Emissions[a] (million metric tons CO_2 equivalent)	% Change 1990–98	Kyoto Protocol Target (% change from 1990)
Australia	423	14.5	8.0
Austria	75	6.5	–13.0
Belgium	136	6.5	–7.5
Bulgaria	157	–46.3	–8.0
Canada	611	13.2	–6.0
Czech Republic	190	–22.2	–8.0
Denmark	70	9.5	–21.0
Finland	75	1.5	0.0
France	554	0.9	0.0
Germany	1,209	–15.6	–21.0
Greece	105	18.1	25.0
Hungary	102	–17.7	–6.0
Ireland	53	19.1	–13.0
Italy	519	4.4	–6.5
Japan	1,175	9.4	–6.0
Netherlands	218	8.4	–6.0
Norway	52	7.7	1.0
Poland	564	–28.7	–6.0
Portugal	64	17.2	27.0
Romania	229	–28.5	–8.0
Russian Federation	2,999	–29.6	0.0
Spain	306	21.0	15.0
Sweden	69	6.4	4.0
Switzerland	53	1.3	–8.0
Ukraine	919	–50.5	0.0
United Kingdom	741	–8.3	–12.5
United States	6,049	11.2	–7.0

NOTES: [a]GHG emissions include CO_2, methane, N_2O, HFCs, PFCs, and SF_6 using IPCC global warming potentials.
SOURCE: IEA 2000d.

Organization for Economic Cooperation and Development (OECD) nations as a whole increased their CO_2 emissions nearly 10 percent between 1990 and 1999 (EIA 2001b). Eastern European nations and the former Soviet Union, on the other hand, cut their emissions 39 percent between 1990 and 1999 due mainly to the severe economic contraction following the collapse of communism there. As a result, so-called Annex I nations (38 industrialized nations) as a whole most likely met the goal of limiting their combined greenhouse gas emissions in 2000 to the level in 1990 (EIA 2001b).

As evidence of global warming and its potentially devastating impacts mounted during the 1990s, nations came together to negotiate the Kyoto Protocol to the Climate Change Convention. The Kyoto Protocol established concrete and binding emissions reduction targets for the Annex I nations starting in the period from 2008 to 2012. The target for all Annex I countries is a 5.2 percent reduction relative to emissions in 1990, with differentiated targets by country and region (see Table 7-3). Given that emissions are well above 1990 levels and rising in OECD nations as a whole, ratifying and having the Kyoto Protocol take effect could spur energy efficiency and renewable energy efforts.

In early 2001, the Bush administration indicated that the United States will not ratify the Kyoto Protocol in spite of the fact that it is responsible for about 30 percent of worldwide CO_2 emissions historically and about 25 percent of current CO_2 emissions. Then the Bush administration issued a national energy plan that would greatly increase U.S. dependence on fossil fuels and CO_2 emissions (NEPDG 2001). It announced its alternative CO_2 emissions reduction strategy in early 2002, but this strategy consisted of further appeal for voluntary action and was widely criticized as little more than business-as-usual (Gardiner and Jacobson 2002, Krugman 2002).

The position of the Bush administration leaves the future of the Kyoto Protocol somewhat in doubt. It is possible for the treaty to take effect if it is ratified by at least 55 nations responsible for at least 55 percent of total greenhouse gas emissions from industrialized countries as of 1990. Most of the operational details of the Kyoto Protocol were worked out in 2001 (see discussion in the following text). Furthermore, the treaty gained political support among industrialized nations other than the United States during this process. The European Union nations and Japan ratified the Kyoto Protocol in mid-2002, meaning the Protocol would take effect even without the United States if it is also ratified by Russia (Claussen 2002). Russia announced its intention to ratify at the World Summit on Sustainable Development in August 2002.

If the Kyoto Protocol enters into effect, it includes a number of so-called flexibility mechanisms that are intended to reduce the cost of compliance and stimulate investment in clean energy technologies in developing nations and economies in transition. The primary objective of these mechanisms is to allow greenhouse gas emissions reductions to take place and emissions reductions targets to be met at the lowest overall cost. However, the flexibility mechanisms limit the environmental effectiveness of the Protocol due to the so-called hot air trading allowance explained in the next section (Agarwal 1999, den Elzen and de Moor 2001).

Joint Implementation

Article 6 of the Kyoto Protocol allows joint implementation (JI) between Annex I nations that have fixed emissions limits under the Protocol. It allows parties investing in such projects to receive transferable emissions reduction credits, with the emissions reductions subtracted from the cap of the countries that sell the credits. This provision allows Western nations to invest in and receive emissions credits from energy efficiency projects in Eastern Europe and the former Soviet Union (Petkova and Baumert 2000). But Western nations can also purchase excess emissions credits that the transition countries will have due to their economic contraction since 1990 (this is known as "hot air trading").

JI could help to stimulate energy efficiency and renewable energy projects in the transition countries, but it remains to be seen how much legitimate JI will take place and what overall effect it will have on clean energy development and efficiency improvement in these nations. Since energy waste is widespread and efficiency improvements are very cost-effective, energy efficiency upgrades in factories and buildings are already being implemented in Eastern Europe and the former Soviet Union (Chandler et al. 1999). It is unclear whether JI will significantly accelerate this positive trend.

The political agreement reached in Bonn, Germany, in 2001 established the operating rules for how JI will work if and when the Kyoto Protocol takes effect. First, it was agreed that there would be no limits on emissions trading, including hot air trading. However, a qualitative requirement states that domestic action shall constitute a "significant element" of the effort by each Annex I country (den Elzen and de Moor 2001). Second, emissions reductions from nuclear energy facilities are not eligible for JI projects or emissions trading. Third, to prevent overselling of emissions credits, countries must retain emissions allowance reserves and are subject to penalties for noncompliance with the treaty during each commitment period. Fourth, governments are given the opportunity to review and approve JI projects within their borders. Since the country will have to make up the "exported" emission reductions, it is reasonable to allow such review and approval.

Clean Development Mechanism

Article 12 of the Kyoto Protocol provides for a Clean Development Mechanism (CDM) that could lead to additional investment in renewable energy, energy efficiency, and natural gas projects in developing countries. The CDM allows parties in Western nations to receive emissions credits for investing in projects that reduce greenhouse gas emissions in non–Annex I countries, as long as the projects advance sustainable development within the host country. According to the Protocol, the CDM can generate certifiable emissions credits prior to the first commitment period of 2008–2012 (unlike emissions trading under JI, which begins in 2008). When CDM begins, it could lead to tens of billions of dollars of investments in clean energy technologies in developing countries over a 10-year period (Thorne and LaRovere 1999).

The CDM will operate in a number of ways, including: (1) allowing bilateral projects whereby Annex I country investors directly participate in CDM projects, (2) multilateral projects whereby Annex I country investors work through a centralized investment fund that could be involved in project development, or (3) unilateral efforts whereby a host country sponsors CDM projects and markets the resulting emissions credits. In order for the CDM to be implemented properly, an effort should be made to exclude projects that likely would have occurred in the absence of CDM emissions credits. If this is not done, the CDM credits could lead to a net increase in global emissions (Bernow et al. 2000).

Some restrictions on CDM projects were agreed to in 2001. First, nuclear energy projects are excluded. Second, carbon sequestration projects are limited to forestation projects during the first commitment period, and the use of credits from sequestration projects is capped at 1 percent of base-year emissions for each Annex I country. These restrictions are reasonable given the complexity, uncertain permanence, and uncertain net impacts of carbon sequestration projects in developing countries (Thorne and LaRovere 1999). A CDM Executive Board was established in 2001 to work out remaining operational details such as procedures for calculating project baselines and project monitoring requirements.

Implications of the U.S. Withdrawal

The withdrawal of the United States from the Kyoto Protocol will have a number of adverse effects on efforts to limit greenhouse gas emissions and hasten a clean energy revolution worldwide. First, it reduces the impetus for increasing energy efficiency and renewable energy use in the United States. Second, it reduces the effectiveness of the Kyoto Protocol, assuming it is ratified and enters into force. One study estimates that the combination of U.S. withdrawal and the flexibility mechanisms will result in Annex I countries other than the United States reducing their emissions by 130 MMT of carbon equivalent per year (3 percent) between 2008 and 2012, compared to a 755 MMT (17 percent) annual reduction with the United States included (den Elzen and de Moor 2001).

The withdrawal of the United States along with allowing hot air trading means that other Western countries could buy a large portion of their required emissions reductions from Russia and other Eastern European countries during the first commitment period, unless they voluntarily agree to limit their exchange of hot air credits. Without such a limit, there would be much less pressure to increase energy efficiency, expand renewable energy use, and implement other low carbon technologies in Western Europe, Japan, and other participating Annex I countries.

This situation could also threaten the development of a meaningful emissions trading market. One study estimates that the carbon credit price in the international market will be cut from about $36 per ton of carbon to $9 per ton because of the withdrawal of the United States and other rules established for the Protocol (den Elzen and de Moor 2001). Developing countries would receive only about $500 million per year from CDM projects during the first commitment period according to this analysis. Furthermore, the focus is likely to be on low-cost credits (in terms of cost per unit of avoided CO_2 emissions) as well as projects with low transaction costs. This could mean little or no additional funding for smaller-scale technologies or projects targeted to rural areas.

Even though the Kyoto Protocol may have relatively limited near-term environmental benefits, an imperfect agreement is better than none. If other nations move ahead with Kyoto Protocol implementation in spite of the U.S. withdrawal, it will increase pressure for the United States to act responsibly and join other nations in limiting the risk of potentially catastrophic climate change. And if Western nations commit to achieving a majority of emissions reductions domestically, they should be able to demonstrate that a shift to a cleaner, more efficient energy future is possible without significant economic hardship.

If the United States reverses its position and participates in the Kyoto Protocol, the consequences for longer-term global energy trends would be positive but still modest. As Eileen Claussen, president of the Pew Center on Global Climate Change has stated,

> [T]he Kyoto Protocol is just a first step on what will be a long march to a less carbon-intensive world. Its initial targets for emissions reductions take us only to the 2008–2012 period, and they represent just a very small down payment on the level of reductions that scientists say

> we must achieve to have a real effect on mitigating climate change. (Claussen 2001)

The Kyoto Protocol points industrialized countries in the direction of lower fossil fuel consumption and CO_2 emissions, and starts them down this path. Much more will need to be done in the coming decades to further reduce CO_2 and other greenhouse gas emissions while fostering economic and social development worldwide.

Enhancing International Technology and Policy Cooperation

A wide range of institutions are engaged in international clean energy cooperation. These institutions are making important contributions to policy innovations, capacity building, and technology transfer. But there are a number of limitations to the current set of activities and institutions.

First, the IEA is not a truly global agency. Developing countries and other nonmembers are not involved in defining the goals or setting the priorities of the IEA. These countries can participate in the implementing agreements, but only a few nonmember countries such as Russia, South Korea, and Mexico routinely do this (IEA 2001h). While promoting greater energy efficiency and alternative energy sources is listed as one of the five principal aims of the IEA, it is not the primary mission and often takes a back seat to other objectives such as oil market coordination, ensuring emergency response capabilities, and energy forecasting. Moreover, the IEA only has about 10 staff and a relatively modest budget devoted to promoting greater energy efficiency and renewable energy use.

The United Nations has energy programs scattered throughout a number of its agencies. These programs are modest in size and at times duplicative or competitive. No individual United Nations agency has a "critical mass" or strong mandate to advance a clean energy transition. As a result, the United Nations Framework Convention on Climate Change has become a de facto global energy forum and a major mechanism for decisions related to capacity building, technology transfer, and clean energy finance worldwide.

The GEF is helping developing countries acquire and build sustainable markets for clean energy technologies. But the GEF has been criticized because of its structure and operating procedures (Ramakrishna and Young 1997). Some developing countries view the GEF as too closely linked to and dominated by the World Bank. They would prefer that the GEF be set up either as an independent organization or as an organization operating under the Climate Change Convention. Also, some developing countries believe the GEF is not giving adequate attention to socioeconomic development.

Bilateral assistance is often self-serving and/or driven by political concerns, for example in promoting exports from donor countries or in selecting countries to support. Bilateral assistance is largely uncoordinated and frequently not responsive to the needs of developing countries. Because of these concerns, new funding mechanisms, complementary to the GEF and ongoing bilateral assistance, were created for fostering technology transfer, greenhouse gas emissions mitigation, and climate change adaptation in developing countries as part of the 2001 agreement concerning implementation of the Kyoto Protocol (den Elzen and de Moor 2001).

In summary, current international efforts supporting a clean energy transition worldwide are fragmented and suffer from a number of shortcomings. No single institution is dedicated to fostering this transition. The multitude of actors often leads to duplication or conflicting efforts. Literally dozens of organizations and programs offer developing countries training or technical assistance related to clean energy technologies and policies, often overwhelming and confusing these countries.[3] This situation suggests that a new consolidated institution is needed—one that is truly global in nature and has a clear mandate to foster a clean energy transition.

An International Energy Efficiency and Renewable Energy Agency

An International Energy Efficiency and Renewable Energy Agency (IEEREA) could be created to support and strengthen energy efficiency and renewable energy efforts in both industrialized and developing countries. It is important that an IEEREA accelerate energy efficiency and renewable energy implementation worldwide so that it is not just another vehicle for the North "preaching" to the South while failing to "practice what it preaches." Also, working globally in a coordinated manner could create economies of scale and help clean energy technologies meet cost and deployment targets.

Implementation of energy efficiency and renewable energy policies and technologies is the responsibility of individual nations and private markets. An IEEREA could support and strengthen national and private sector initiatives through technology and policy cooperation, capacity building, and the like. The IEEREA could also serve as a forum for discussing and negotiating global energy efficiency and renewable energy targets on an ongoing basis.[4] But it would not substitute for national, state, and local action. Appropriate activities for the IEREA could include:

- preparing technology and policy reviews, databases, product directories, and resource assessments;
- disseminating and exchanging information;
- facilitating collaborative RD&D;
- offering training courses and supporting national or regional training programs;
- establishing uniform testing and rating procedures;
- encouraging international trade and joint ventures in clean energy technologies;
- developing overall energy efficiency and renewable energy implementation targets;
- setting cost and performance goals for individual clean energy technologies;
- developing multilateral or coordinated energy tax, pricing, or procurement initiatives;
- encouraging financing initiatives and assisting existing financial institutions such as the World Bank and other MDBs;
- assisting governments who request help in developing or analyzing policy instruments;
- supporting national or regional energy efficiency and renewable energy agencies or centers;
- recognizing outstanding actions in clean energy technology development and deployment by governments, private companies, and nongovernmental organizations; and
- developing and disseminating analytical, modeling, and evaluation tools.

If an IEEREA were formed, it could combine and build on the efforts of the large number of bilateral and multilateral organizations working on clean energy development today.[5] One option would be to split off and consolidate the clean energy activities of the International Energy Agency and United Nations agencies. It may be feasible and desirable to fold the clean energy activities of the Global Environmental Facility into the IEEREA as well. Furthermore, an IEEREA could be assigned a primary role for clean energy technology cooperation under the United Nations Framework Convention on Climate Change. In addition, individual countries could be encouraged to dedicate some or all of their energy assistance resources to the Agency.

While it is difficult to envision creating a new international energy institution, it has been done before. The International Atomic Energy Agency (IAEA) was created during the late 1950s as a specialized agency within the United Nations system to support nuclear safety, monitor the distribution of nuclear fuels, and promote nuclear energy development and technology transfer. The IAEA has 128 member nations and a budget of about $230 million per year. It carries out many of the activities suggested here, but for nuclear energy rather than energy efficiency and renewable energy (PCAST 1999).

Various international committees over the past 20 years have recommended creating an international renewable energy agency, but these proposals were blocked by the United States, Japan, and other countries (Eurosolar 2001). Today it might be possible to establish an IEEREA under the auspices of the Framework Convention on Climate Change and Kyoto Protocol to the Convention, especially if it meant consolidating and enhancing current disparate efforts.

Initial funding and staff for the IEEREA could come from combining some or all of the programs listed here. Additional funding could be provided by national governments or by earmarking a portion of energy or carbon taxes enacted by various industrialized countries. Adopting a small carbon tax in all OECD nations could generate substantial new resources (e.g., a tax of just $0.10 per metric ton of carbon would generate about $400 million per year). An IEEREA could also be funded by having nations pledge a portion of savings to governments from reducing subsidies for conventional fossil fuels and nuclear energy.

Once it reaches maturity, the IEEREA might spend $1 to $2 billion per year to carry out all of the functions suggested here. For comparison, the U.S. Department of Energy was spending approximately $1.2 billion per year for energy efficiency and renewable energy RD&D and deployment programs as of 2002. While an annual budget of as much as $2 billion may sound like a lot of money, it is on the order of one-tenth of 1 percent of current energy expenditures worldwide and on the order of 1 percent of desired investment in energy efficiency and renewable energy technologies worldwide.

A well-funded IEEREA would need a headquarters, but it could also have branches on each continent. These branches could coordinate activities in their region, working closely with individual countries. The branches could also host "Centers of Excellence" in particular technologies, research areas, or policy approaches (e.g., a bioenergy center or an efficiency labeling and standards center). While many activities could be decentralized, it would be logical to conduct some activities such as organizing collaborative RD&D, developing uniform test procedures, setting performance targets for "global products," and data collection and analysis centrally.

Notes

1. Smelt reduction is an advanced process for integrated steelmaking that is both more energy efficient and less polluting than conventional processes.
2. The regional development banks include the African, Asian, and Inter-American Development Banks and the European Bank for Reconstruction and Development.
3. I experienced this situation firsthand when I worked for the PROCEL energy efficiency program in Brazil in 1996–97.
4. Nations attempted but failed to establish a global renewable energy target at the August 2002 Johannesburg Summit on Sustainable Development. Having a forum to debate and negotiate such targets could help to avoid this type of failure in the future.
5. To be clear, I am not proposing yet another international energy program or institution that would add to the fragmentation and confusion that exists today. I am suggesting integrating and expanding current efforts.

Toward a Sustainable Energy Future

An energy inefficient and fossil fuel–intensive energy future would present a variety of problems for the world including:

- rapid global warming,
- high levels of investment in energy supply,
- increased local and regional air pollution,
- greater national and international security risks,
- rapid petroleum depletion, and
- continued inequity.

Together, these amount to a serious threat to the environmental integrity of our planet, our standards of living, and the ability of the developing world to climb out of poverty. A business-as-usual energy future is neither sustainable nor desirable.

Fortunately an inefficient, fossil fuel–intensive energy future is not inevitable. By emphasizing increased energy efficiency, greater reliance on renewable energy sources, and greater use of natural gas for a number of decades, all of these problems can be mitigated. In short, a "revolution" in the way the world produces and consumes energy is possible and would provide a wide range of economic, environmental, and social benefits. By emphasizing cost-effective energy efficiency improvements along with a shift to renewable energy sources, nations of the world need not sacrifice economic growth to protect the environment and provide modern energy sources to the roughly 2 billion people now lacking them.

However, the barriers that limit the introduction and deployment of energy efficiency, renewable energy, and natural gas technologies in both developing and industrialized nations are both real and formidable. These barriers pertain to technology availability and performance, consumer and business decision making, market organization, energy prices and taxes, regulations, and political forces. Indeed various business-as-usual energy forecasts show that without substantial policy reforms, the United States and other countries will continue to make modest energy efficiency improvements and will continue to rely heavily on fossil fuels, particularly petroleum and coal, which are the most problematic fuels.

What then are the overall prospects for an energy revolution? I have argued that an energy revolution is possible given the combination of technological advances and policy experience over the past 30 years. It is useful to review the lessons from this experience before considering how a global energy revolution might unfold.

Policy Lessons

The most important lesson that can be drawn from past experience is that public policies that are well-designed and implemented can overcome the barriers to greater energy efficiency, renewable energy use, and cleaner fossil fuel technologies. This conclusion was confirmed in the case studies of policy initiatives worldwide as well as in forward-looking policy assessments for the United States and Brazil.

There are also a number of more specific lessons that can help guide a revolution in the way the world produces and uses energy. What should we do?

- Aim to transform markets. Integrate policies into market transformation strategies, addressing the range of barriers that are present in a particular locale. Make the policies strong enough to remove or overcome the barriers that are present. And allow them to evolve over time as some barriers are removed and others come to the forefront.

- Create an "innovation system" whereby clean energy technologies and services become established in the market, markets develop, and costs decline as market share grows. This approach often involves a combination of efforts focused on both technology development and product supply on the one hand and consumer demand and market development on the other.
- Make policies predictable and stable to reduce the risk and uncertainty that investors, businesses, and consumers face. Keep them in place for a decade or more to ensure an orderly development of energy efficiency and renewable energy industries and markets.
- Expand government-funded research, development, and demonstration (RD&D) on clean energy technologies to reduce their cost and improve their performance. Also, expand RD&D on behavioral and implementation-related issues. Foster collaboration between research institutes and the private sector, and combine RD&D with market development efforts.
- Provide convenient financing to increase the adoption of energy efficiency and renewable energy technologies, especially in developing countries. Financing tends to be more effective if loans are provided at attractive terms through existing financial institutions, and if loans are linked to marketing of energy efficiency and renewable energy measures.
- Provide financial incentives to increase the adoption of energy efficiency and renewable energy measures. Financial incentives should reward superior performance (e.g., pay for energy savings or renewable energy production). Also, incentives should diminish or phase out as markets for energy efficiency and renewable energy measures expand and their costs drop.
- Reform energy prices. Eliminate subsidies for fossil fuels and enact taxes based on their environmental and social costs. Use some of the tax revenue to support energy efficiency and renewable energy initiatives to maximize the energy, environmental, and economic benefits.
- Enact regulations or market obligations to stimulate widespread adoption of energy efficiency improvements or renewable energy sources. Make sure these regulations or obligations are technically and economically feasible, enforce them, and update them periodically. Also, structure emissions cap and trading schemes so that they encourage and provide credit for emissions reductions from end-use efficiency improvements and renewable energy technologies.
- Adopt voluntary agreements between governments and the private

sector in situations where regulations or market obligations cannot be enacted or enforced. Complement voluntary agreements with financial incentives, technical assistance where needed, and the threat of taxes or regulations should the private sector not meet their commitments.

- Introduce greater competition to achieve efficiency improvements, cost reductions, and reduced emissions in electricity supply. But adopt renewable energy and energy efficiency obligations or funding mechanisms in conjunction with these reforms to foster rather than hinder renewable energy and end-use efficiency efforts.
- Disseminate information and provide training to increase awareness and improve know-how with respect to energy management and renewable energy options. Combine these efforts with incentives, voluntary agreements, or regulations to increase their impact.
- Use bulk procurement to help commercialize and establish initial markets for innovative clean energy technologies. Governments should purchase energy-efficient products, renewable energy devices, or "green power" for their own use, as well as sponsor and help organize bulk purchases by a wide range of public and private entities.
- Build capacity to implement effective energy efficiency and renewable energy policies and programs in all countries. Also, train and support businesses that manufacture, market, install, and service clean energy technologies.
- Carry out both integrated energy resource planning and integrated transportation and land use planning to guide investments to options that minimize overall societal costs (including environmental costs). Energy and transportation plans should contain concrete goals, actions for achieving the goals, and monitoring and evaluation procedures.

To have a large-scale and lasting impact, energy policies must engage the private sector in production, marketing, and adoption of energy efficiency, renewable energy, and other low emissions technologies. Without strong private sector actors, energy efficiency and renewable energy efforts are likely to be ad hoc, limited in scope, and limited in duration (Kammen 1999, Martinot 2001). Likewise, investments must be drawn away from inefficient, polluting energy technologies and toward clean energy technologies. Governments should create a market environment where investors and private companies innovate, compete, and ultimately profit from investments in clean energy technologies.

Clearly some energy companies have a vested interested in maintaining an inefficient and carbon-intensive energy future. But many other companies are actively developing, producing, or investing in high-efficiency, renewable energy, and advanced fossil fuel technologies. Policy makers can work with these progressive companies to develop and implement policies and programs that move nations or regions toward a more sustainable energy future.

In general, policies to advance energy efficiency and renewable energy use will be more effective if backed by a high-level government commitment. As shown in Chapter 4, government backing played an important role in Brazil's ethanol fuel program, China's national energy efficiency and improved cookstoves efforts, Denmark's wind power program, India's renewable energy efforts, the Dutch industrial voluntary agreements, and California's energy efficiency efforts. A high-level commitment will help to sustain policies and programs over the long run, provide legitimacy for new technologies, and encourage investment in clean energy technologies by the private sector.

For the most part, this book examined policies that were successful in advancing energy efficiency, renewable energy sources, or natural gas use. But there have been many failed policy initiatives aimed at promoting clean energy technologies. For example, many nations have not succeeded in establishing viable markets for renewable energy technologies in spite of considerable RD&D or incentive programs (Johnson and Jacobsson 2001, Martinot et al. 2002). Other nations have had little success with efforts to increase the efficiency of passenger vehicles over the past 10 to 15 years.

Fortunately, we can identify the reasons for most of these failures. Failed policy initiatives usually lack some if not many of the characteristics of successful efforts: they do not remove or overcome all of the major barriers, they are not part of an integrated market transformation strategy, they lack continuity or a high-level government commitment, they do not engage the private sector, and/or they do not develop a favorable market environment. While it is possible to overcome the barriers and implement energy efficiency, renewable energy, and clean fossil fuel technologies on a large scale, policies must be carefully designed and implemented.

Chapter 7 addressed the international dimension of advancing toward a more sustainable future. The threat of global warming has increased international cooperation related to clean energy development and deployment during the past decade. Implementation of the United Nations Framework Convention on Climate Change has contributed to capacity building, technology transfer, and clean energy finance for developing countries. At the same time, many industrialized nations have expanded their cooperative energy efficiency and renewable energy efforts at the regional or international level, often in conjunction with commitments to cut greenhouse gases.

International institutions including the International Energy Agency (IEA),

Global Environmental Facility (GEF), World Bank, United Nations, and bilateral assistance agencies are engaged in many useful clean energy initiatives. But current international efforts are fragmented, limited in scale, and in some cases self-serving or conflicting. No international agency has a clear mandate or adequate resources to lead a clean energy revolution. As explained in Chapter 7, this situation suggests that a new International Energy Efficiency and Renewable Energy Agency is needed. The new agency could be formed by consolidating the clean energy efforts of various other agencies such as the IEA, United Nations agencies, GEF, and bilateral donors. The agency could support national, regional, and private sector efforts by facilitating collaborative RD&D; tax, pricing, or procurement initiatives; capacity building; international trade and joint ventures; harmonization of testing procedures and efficiency standards; setting of energy efficiency and renewable energy implementation targets; and the like.

Progress to Date

As noted in Chapter 1, modern renewable energy sources including hydropower now provide close to 5 percent of global energy supply. Global energy use is rising about 2 percent per year in spite of energy efficiency improvements in most countries. There will be large increases in oil and coal use over the next few decades unless energy efficiency improvements and renewable energy production accelerate (IEA 2000a). Whether energy efficiency and renewable energy can accelerate depends in part on where they stand today.

Wind power is the world's fastest-growing major energy source. Figure 8-1 shows the growth of wind power capacity worldwide over the past 15 years. Cumulative installed wind capacity increased by more than a factor of 10 from 1992 to 2001. The 6,500 megawatts (MW) of capacity installed in 2001 alone represented about $7 billion of investment in new wind turbines (AWEA 2002). With this steady growth, wind power provided about 60 terawatt-hours (TWh) or 0.4 percent of the electricity produced worldwide in 2001. Furthermore, wind power capacity is now increasing 30 to 35 percent per year. This high rate of growth is expected to continue at least through 2005 due to technological and market advances, the cost effectiveness of modern wind turbines, and renewable energy policy initiatives in many countries (BTM Consult 2001).

Figure 8-2 shows the growth in photovoltaic (PV) capacity worldwide over the past 15 years. Cumulative installed capacity increased by a factor of 5 between 1992 and 2001. Installed capacity increased about 25 percent per year in 2000 and 2001 due in large part to the grid-connected PV programs in Japan and

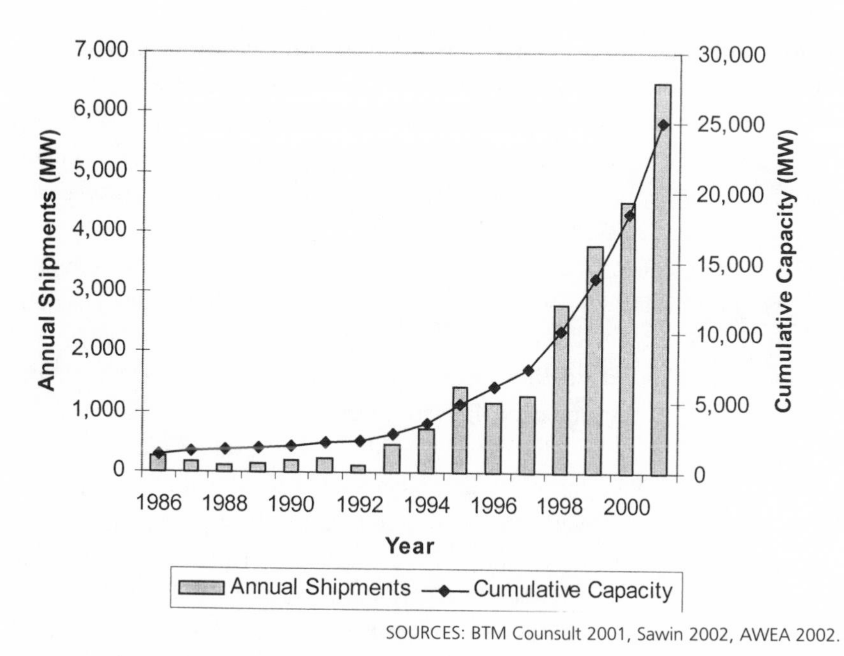

SOURCES: BTM Counsult 2001, Sawin 2002, AWEA 2002.

FIGURE 8-1: Growth in global wind power capacity.

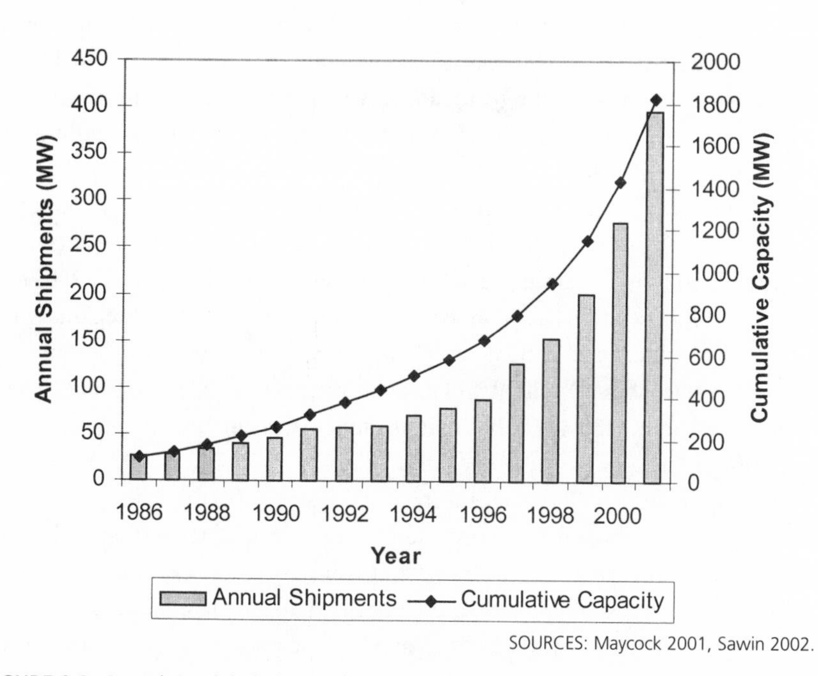

SOURCES: Maycock 2001, Sawin 2002.

FIGURE 8-2: Growth in global photovoltaic power capacity.

Germany. However, cumulative PV capacity was only about 1,800 MW as of 2001, meaning solar power systems provided about 3.6 TWh or 0.025 percent of the electricity produced worldwide in 2001. But with incentive programs expanding in many countries, new solar products under development, and PV costs falling, grid-connected PV capacity alone could expand 35 percent per year between 2001 and 2005 (Rever 2001).

Comparable data on global production and use of modern forms of bioenergy are not available. The growth rates have been relatively high in some countries such as Brazil and Scandinavian nations, but more modest elsewhere. For example, it is estimated that the supply of all forms of bioenergy, including waste materials from the pulp and paper industry, municipal solid waste that is burned for energy, fuelwood, and ethanol fuel roughly doubled in the United States between 1971 and 2000 (Overend 2002). Bioenergy is now the leading renewable energy source in the United States, accounting for about 3.5 percent of total energy supply as of 2000. In 1999, the federal government adopted a goal of tripling the use of bioenergy and biobased products in the United States by 2010 (Wright and Kszos 1999). Bioenergy sources already provide 10 to 20 percent of the total energy supply in some European countries including Austria, Finland, and Sweden (IEA 2001e).

The adoption of energy-efficient technologies has also grown rapidly during the past decade, although worldwide sales data are not readily available for most devices. Compact fluorescent lamps (CFLs), however, are one product for which sales estimates are available. Annual CFL sales increased by about a factor of 5 between 1992 and 2001 (Fig. 8-3). Sales are now increasing 25 to 30 percent per year due to incentive and promotion programs in many countries, combined with product improvements and cost reductions in recent years. CFL production is expanding particularly rapidly in China (Scholand 2002). In the United States, CFL sales are booming mainly due to state and regional promotional and incentive campaigns (Calwell et al. 2002). Assuming a four-year life on average, about 2 billion CFLs were in use worldwide as of 2001. These lamps saved about 135 TWh that year, meaning they cut global electricity consumption in 2001 by about 1 percent.

It is clear that there has been substantial progress with the development and adoption of key renewable energy and energy efficiency technologies. But we have much further to go. In this respect, it is instructive to examine what leading countries, states, or regions are committed to or planning. Also, longer-term sales and market forecasts are available for some of the clean energy technologies.

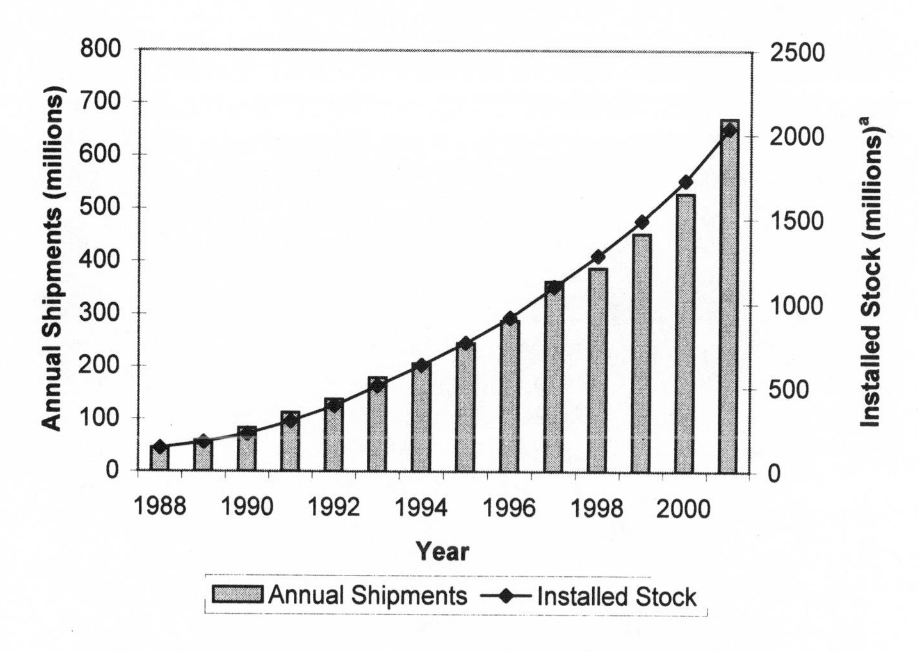

NOTE: [a]Assuming lamps have a four-year life on average.
SOURCES: Borg 2002, author's estimate for 2001.

FIGURE 8-3: Growth in global use of compact fluorescent lamps.

Future Prospects

The European Union aims to double its renewable energy use by 2010, meaning that renewable sources would provide 12 percent of total energy supply and 22 percent of electricity supply that year. Some European countries have adopted their own more ambitious goals. For example, the United Kingdom plans to increase the fraction of electricity generated by renewable energy from 2 percent in 2000 to about 10 percent in 2010 (IEA 2001e). Spain plans to obtain 12 percent of its electricity in 2010 from wind power alone (BTM Consult 2001). Meanwhile, Denmark plans to increase the contribution of renewable energy sources from about 10 percent of total energy supply in 2000 to 18 percent by 2010 and 35 percent by 2030 (Odgaard 2000).

In the United States, some states and municipalities are providing leadership on energy efficiency and renewable energy implementation. California has adopted new appliance efficiency standards, tougher building energy codes, and expanded state and utility energy efficiency programs to further reduce its energy intensity, which as noted in Chapter 4 is already low compared to that of other states. California has also adopted the goal of increasing the contribution of renewable energy to total electricity supply from 12 percent in 2001 to 17 percent by 2006 and has set up renewable energy incentive programs to achieve this goal (CEC 2001b). In addition, California enacted legislation in 2002 that directs state officials to establish CO_2 emissions standards on new cars and light trucks starting in 2009 (Hakim 2002). This action, which is vigorously opposed by most of the auto industry, could end the stalemate on new vehicle fuel efficiency in the United States.

California is not the only state that has made a strong commitment to energy efficiency and renewable sources. New York has adopted the goals of cutting primary energy use per unit of economic output 25 percent by 2010 (relative to the level in 1990), increasing renewable energy use 50 percent by 2020, and reducing greenhouse gas emissions 5 percent below 1990 levels by 2010 (NYSEPB 2002). New Jersey has expanded its energy efficiency and renewable energy programs and is pursuing voluntary commitments to reduce statewide carbon dioxide (CO_2) emissions by 13 percent compared to otherwise projected levels in 2005 (Fialka 2001). Nevada is requiring that 15 percent of its electricity come from renewable sources by 2013. Texas added around 900 MW of renewable capacity in 2001 alone, nearly half the goal it set for 2009. And New Hampshire has adopted stringent emissions standards, including standards on carbon dioxide emissions, for its fossil fuel–burning power plants.

Some developing countries have also made significant commitments to efficiency and renewable energy sources. As discussed in Chapter 6, Brazil already obtains over half its new energy supply from renewable sources. India plans to obtain 10 percent of its incremental electricity generation from renewable sources by 2012, building on its extensive policies and significant achievements to date (Timilsina, Lefevere, and Uddin 2001). China plans to add 20,000 MW of renewable electricity capacity (excluding large-scale hydropower) by 2010 (Martinot et al. 2002). And Thailand plans to obtain one-third of its energy needs from energy efficiency and renewable energy sources by 2011 (NEPO 2002). Thailand is implementing a variety of energy efficiency and renewable energy programs to achieve this goal.

Future projections for some specific technologies can also help to inform how a clean energy revolution might unfold. It is conceivable that global wind power capacity could expand 20 to 25 percent per year over the next two decades if projected cost and performance improvements occur, and there is adequate policy support (EWEA, Forum for Energy and Development, and Greenpeace International 1999, Williams 2002). If capacity expands 25 percent per year between 2002 and 2010 and 20 percent between 2011 and 2020, there would be a total of 1,200 gigawatts (GW) of wind power capacity in the world by 2020. This amount of capacity would generate approximately 3,200 TWh of electricity annually, given anticipated future capacity factors—equivalent to about 18 percent of the world's total electricity supply in 2020 assuming energy efficiency efforts limit electricity demand growth to about 1 percent per year.

If wind capacity expands to this degree, the cost of wind power is projected to fall to $0.03 per kWh or less by 2020 as a result of learning effects and technological advances (Turkenburg 2000, Williams 2002). Reaching this cost level would facilitate further implementation including transmitting wind power over long distances (e.g., transmitting wind power from the Great Plains to more heavily populated regions of the United States and from Inner Mongolia to more heavily populated regions of China) (Williams 2002). Wind power could supply up to half the world's electricity by 2040 assuming output continued to grow 6 to 7 percent per year after 2020.

Regarding solar PV power, it is conceivable that installed capacity worldwide could expand 30 percent per year over the next two decades if strong supporting policies are adopted in many more countries (Cameron et al. 2001, Williams 2002). Growth at this rate would lead to about 260 GW of solar capacity by 2020. This amount of PV capacity would generate approximately 520 TWh of electricity per year—equivalent to 3 percent of the world's total electricity supply in 2020 assuming 1 percent per year demand growth.

Solar power would represent a small but important share of total global electricity supply in 2020 if this target can be met. It would enable basic electrification of a large fraction of the roughly 500 million households in developing countries that do not use electricity today.[1] Nevertheless, most of the installed PV capacity in 2020 is likely to be grid-connected. And once again, cost reductions due to a combination of expected technological advances, construction of larger production facilities, and learning effects would position solar PV power to further expand in the post-2020 time frame (Williams 2002). If PV capacity continues to expand 10 to 15 percent per year after 2020, solar PV power could supply over 7,000 TWh by 2040 (Cameron et al. 2001). If this level of electricity production were achieved, solar PV systems could contribute up to one-third global electricity supply in 2040, again assuming demand growth is limited to 1 percent per year.

A Global Clean Energy Scenario

While these forecasts for individual technologies are promising, they fall short of what we might consider to be a true energy revolution. What could happen if comprehensive policies along the lines suggested in this book are widely adopted in the coming decades? It is impossible to know for certain. However, scenarios can be created based on assumptions that are consistent with widespread adoption of such policies as well as the forecasts for individual clean energy technologies cited above.

The "ecologically driven" energy scenario presented in Chapter 1 is one example of what is possible. This scenario was developed by the International Institute for Applied Systems Analysis (IIASA) and the World Energy Council (WEC). It assumes expanded national and international efforts support clean energy development throughout the twenty-first century. In it, widespread energy efficiency improvements limit growth in energy use worldwide to about 0.8 percent per year on average. Renewable energy sources provide 40 percent of total global energy supply by 2050 and 80 percent by 2100 (see Fig. 1-5). Solar energy and modern forms of biomass are the two dominant energy sources in the latter part of the century. Worldwide oil use falls by about one-quarter and coal use by one-third by 2050. By 2100, oil and coal together provide only 10 percent of total global energy supply. In addition, there is greater emphasis on improved standards of living in developing countries in this scenario compared to a business-as-usual energy future (Nakicenovic, Grubler, and McDonald 1998).

The ecologically driven scenario is by no means an upper bound on the rate at which efficiency improvements could occur or renewable energy sources could be deployed during the twenty-first century. Table 8-1 and Figure 8-4 show the author's own Global Clean Energy Scenario, illustrating how energy supply and demand could unfold during this century if a strong and steady commitment is made to greater energy efficiency and expanded renewable energy use. This scenario is not derived from assumptions about policy adoption in different

TABLE 8-1

Global Clean Energy Scenario

	Primary Energy (billion tons of oil equivalent)							
Region	1997	2002	2010	2020	2040	2060	2080	2100
OECD	4.88	5.1	4.90	4.66	4.22	3.81	3.45	3.12
EE/FSU	1.03	1.1	1.14	1.20	1.33	1.47	1.62	1.79
LDCs	3.74	3.9	4.29	4.83	6.14	7.79	9.89	12.55
TOTAL	9.65	10.10	10.33	10.70	11.68	13.07	14.96	17.47
Renewables fraction (%)	14	15	18.5	24	40	59	78	100

NOTES: EE/FSU = Eastern Europe and Former Soviet Union; LDCs = less-developed countries; OECD = Organization for Economic Cooperation and Development.

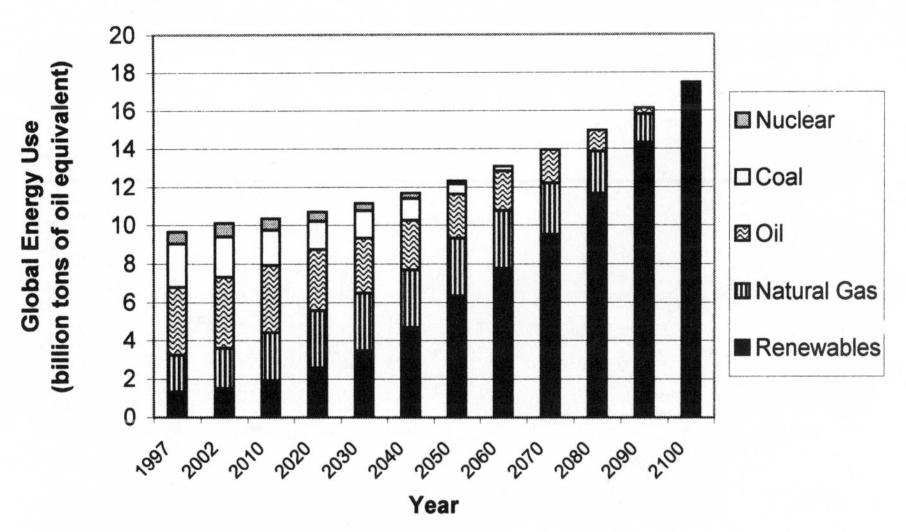

FIGURE 8-4: Global Clean Energy Scenario.

countries and regions. Rather it is based on a self-consistent set of assumptions concerning economic growth, energy intensity reduction, growth in renewable energy supply, and evolution of fossil fuels and nuclear power (see the Appendix to this chapter for details).

Energy efficiency improvements and structural shifts in the Global Clean Energy Scenario limit growth in energy demand during the century to about 0.6 percent per year. Renewable energy supply increases about 2.5 percent per year on average. Renewable sources contribute nearly one-quarter of total global energy supply by 2020, over half by 2050, and all energy supply by 2100. Nuclear energy is phased out within 50 years, coal use in about 60 years, and oil use in about 90 years. Natural gas use increases in the near term and makes an important contribution to energy supplies throughout the century. But natural gas use starts to decline around 2060 and is phased out by 2100.

The Global Clean Energy Scenario would address all of the challenges presented by a business-as-usual, fossil fuel–intensive energy future. Concerning climate change, the Clean Energy Scenario would immediately and steadily reduce CO_2 emissions from burning of fossil fuels, as shown in Figure 8-5. Global emissions would decline 13 percent by 2030, 35 percent by 2050, and 66 percent by 2080 relative to estimated emissions in 2002. The average rate of decline in absolute CO_2 emissions is about 1 percent per year between 2002 and 2050. The decline in CO_2 emissions per unit of GDP is 3.1 percent per annum, twice the rate of decline experienced worldwide between 1971 and 1997 (IEA 2000a). The

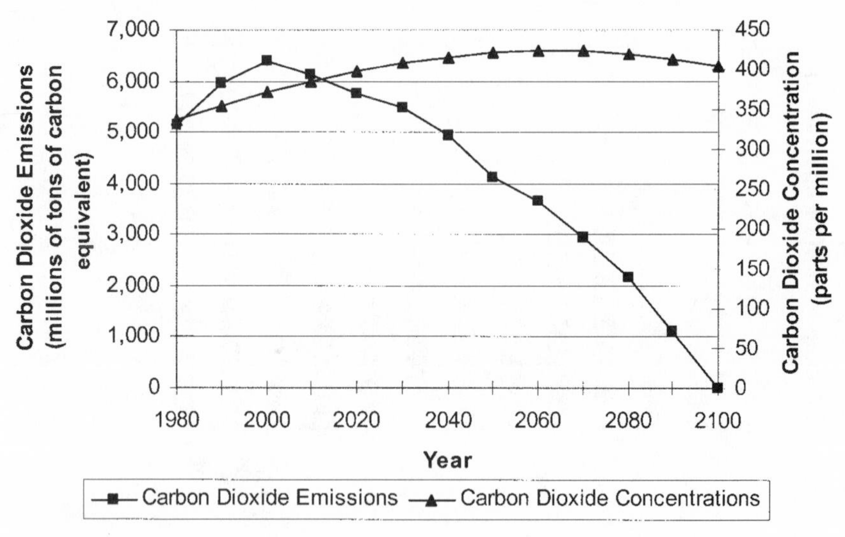

FIGURE 8-5: World carbon dioxide emissions and atmospheric concentrations in the Global Clean Energy Scenario.

rate of decline accelerates in the latter part of the century as fossil fuels are phased out.

If CO_2 emissions steadily decline in this manner, cumulative emissions during the twenty-first century would be limited to about 390 billion tons of carbon equivalent, only about 26 percent more than the roughly 310 billion tons released between 1860 and 2000. Furthermore, this CO_2 emissions trajectory would limit the maximum atmospheric concentration of CO_2 to about 425 parts per million by volume (ppmv), about 52 percent greater than the preindustrialization level of 280 ppmv (see Fig. 8-5). The maximum concentration would occur around the decade from 2060 to 2070 followed by a gradual decline in the latter part of the century. This in turn would limit the increase in global mean temperature since preindustrial times to a range of 0.8 to 2.1°C, with a "best guess" increase of 1.4°C, by 2100.[2] Although there is much uncertainty concerning a "safe" level of temperature increase, various organizations and experts have argued for an upper limit of around 2°C (Schneider and Azar 2001). While the amount of global warming depends on many factors, steadily reducing CO_2 emissions as envisioned in the Global Clean Energy Scenario would greatly reduce the risk of catastrophic climate change. In addition, it may be possible to further reduce the concentration of CO_2 in the atmosphere and corresponding temperature rise if necessary through carbon sequestration (Box 8-1).

Second, the steady decline and eventual phase-out of coal and oil use would

BOX 8.1

What About Carbon Sequestration?

There is considerable interest in capturing and storing CO_2 either in vegetation, the ocean, or geological (underground) sites in order to mitigate human-induced climate change. These options collectively are known as carbon storage or carbon sequestration. The estimated global potential for biological storage through increased forest cover, improved agricultural practices, and the like is on the order of 100 billion tons of carbon equivalent by 2050, although there is substantial uncertainty regarding this potential (IPCC 2001b). Most of the potential is in tropical and subtropical regions. Afforestation and other types of biological carbon storage cost relatively little per unit of carbon abated and can provide other benefits, such as soil and watershed protection, improved biodiversity, and rural employment (Sathaye and Ravindranath 1998). However, realization of this potential depends on land and water availability, widespread adoption of improved land management practices, and other factors. Also, biological storage can be reversed by events such as fire, insect infestation and land use change. Moreover, high rates of biological carbon sequestration cannot be maintained indefinitely.

Another alternative is to capture the CO_2 emitted from fossil fuel-based power plants and other energy conversion facilities, transport it, and dispose of it in the ocean or underground in secure locations. But injection of CO_2 in the deep ocean poses serious ecological risks. Also, some of the CO_2 will eventually escape back to the atmosphere. Therefore, ocean disposal is not considered very appealing and attention is shifting to geological disposal (Williams 2000, Doniger et al. 2002).

Geological disposal is possible in depleted oil and gas reservoirs, coal beds, or in deep saline aquifers. Some underground disposal of CO_2 is already occurring for enhanced oil recovery as well as for environmental reasons (Herzog, Eliasson, and Kaarstad 2000). The barriers to CO_2 sequestration on a large scale in geological formations include the high cost of capturing, transporting, and disposing CO_2 at the present time, as well as immature disposal technologies and uncertain storage capacity, safety, and integrity (Holloway 2001). Considerable RD&D is needed to develop CO_2 capture and disposal techniques, narrow the technical uncertainties, and reduce costs. While CO_2 capture and underground disposal may eventually contribute to global warming mitigation, it is not yet clear if this will be an effective, economical, and acceptable strategy. RD&D on carbon sequestration should continue, particularly on innovative approaches for electricity and fuel production that enable both CO_2 capture and disposal and near-zero emissions of other air pollutants (Williams 2000). But the possibility of carbon sequestration in the future should not stand in the way of accelerated energy efficiency improvements and renewable energy implementation worldwide—that is, a clean energy revolution.

◉ ◉ ◉

significantly reduce local and regional air pollution throughout the world, thereby improving human health and mitigating the adverse impacts of air pollution on agriculture, forests, and other ecosystems. In addition, the emphasis on greater efficiency and use of modern renewable and other energy sources in developing countries would greatly improve indoor air quality and reduce burdensome fuelwood collection, thereby improving health and living conditions, especially for women and children. These effects, observed in the Institute for Applied Systems Analysis–World Energy Council (IIASA–WEC) ecologically driven energy scenario, would be amplified in the Global Clean Energy Scenario, which assumes even higher rates of efficiency improvement and renewable energy growth.

Third, the Global Clean Energy Scenario would lower the total cost to society associated with energy supply and use. While renewable energy technologies and accelerated use of natural gas would have a market cost on par or in some cases slightly greater than conventional energy technologies at least in the near term, the economic benefits of efficiency improvements would more than offset these incremental costs (see Chapter 5). Furthermore, the cost of renewable energy technologies will decline over time through learning effects and technological progress. In addition, the shift away from oil and coal will reduce the environmental and national security costs that are not included in market energy prices but are costs to society nonetheless. And declining demand for fossil fuels should lower their market prices relative to the prices that are expected if demand rises, thereby further benefiting consumers and most businesses.

Fourth, the Global Clean Energy Scenario would reduce oil import dependence and enhance national security for the large majority of nations that are oil importers. In particular it would anticipate and avoid the potential "crunch" if world oil demand continues to rise but conventional oil production peaks at some point in the next few decades. Also, by phasing out nuclear energy, the Clean Energy Scenario reduces the risk of radiation release by accident, terrorist attack on nuclear facilities, or diversion of nuclear materials to weapons.

Although it is not explicitly addressed here, the Global Clean Energy Scenario would help to reduce inequity among nations. Energy consumption declines in Organization for Economic Cooperation and Development (OECD) nations while increasing in developing countries, although energy efficiency improves everywhere. Developing countries account for nearly half of global energy use by 2030 compared to less than 40 percent today (including both modern and traditional energy sources). Developing countries account for 60 percent of global energy use in 2060 and 72 percent in 2100 in the Global Clean Energy Scenario. Furthermore, by emphasizing the adoption of modern renewable energy sources, this scenario is compatible with high rates of social and economic development in the poorer regions of the developing world. Rural areas in particular would benefit from the supply and utilization of energy efficiency measures and renewable energy sources. Per capita energy use in 2100 in industrialized nations would still be higher than in developing nations in the Global Clean Energy Scenario, but the gap between the two major groups of nations would be substantially narrowed.

While it would produce many benefits, a transition from fossil fuels to higher-efficiency and renewable energy sources will not be without its challenges, and may even have some negative effects, albeit limited ones. The storage of energy is one of the technical challenges in a world that runs mostly or entirely on renewable energy sources. Gaseous and liquid fuels produced from renewable sources, such as ethanol and hydrogen, can be stored and distributed in a manner similar to that of natural gas and petroleum products today. But given the intermittent nature of wind and solar energy, new energy storage systems as well as backup power sources will be needed as electricity supply transitions to renewable energy sources. Reservoir-based hydropower where available can enable a large penetration of intermittent renewable sources. In locations where hydropower is not available, it is possible that compressed air storage could be implemented at a modest cost of around $0.01 per kWh (Turkenburg 2000).

The transition from fossil fuels to high-efficiency and renewable energy sources would produce economic "winners" and "losers." The winners should far exceed the losers since energy efficiency and renewable energy measures are much more decentralized and widely distributed than fossil fuels (IEA 2001h). Most renewable energy consumption is likely to be produced domestically rather than imported (Flavin and Dunn 1999). Also, energy efficiency and renewable energy measures tend to support more jobs per unit of energy produced or saved than conventional fossil fuel resources. But the plight of economic "losers" should not be ignored. This is important both as a matter of fairness and as a way to reduce political opposition to the transition.

With respect to potential losers, oil exporting nations and coal producing regions may need some compensation for leaving their fossil fuel resources in the ground. Such compensation could be used in part to develop alternative industries in these nations and regions, including energy efficiency and renewable energy industries. In the United States, for example, bioenergy industries could be developed in coal mining regions of Appalachia, wind power projects and industries in the western states that are heavy fossil fuel producers today, and solar energy industries in the "oil patch." Likewise, solar energy industries and possibly solar-based hydrogen production could be developed in the Persian Gulf region. As these new energy industries take root, special efforts could be made to retrain and hire workers displaced from the fossil fuel industries.

Most of the environmental challenges to a clean energy transition are linked to the characteristics of individual renewable energy technologies. Bioenergy production can contribute to deforestation if it is not produced on a sustainable basis. Bioenergy production can also contribute to land degradation or loss of biological diversity if it is not carefully located and well-managed, and it can contribute to air pollution if biofuels are burned without pollution controls. Bioenergy sources need not have these adverse impacts, however, if they are grown on degraded or excess agricultural lands, and if production and conversion meet strict environmental standards (Turkenburg 2000).

Expansion of hydropower can flood sizable areas, diminish fish stocks in rivers, harm downstream areas, and displace significant numbers of people. But improved technologies, thoughtful dam design, and better water management can reduce these negative impacts. Nevertheless, social and environmental impacts need to be carefully considered before new hydroelectric projects are constructed. These issues no doubt will limit hydroelectric expansion in some countries. Likewise, some countries will choose to restrict natural gas production, for example in wilderness areas. As long as vigorous efficiency improvements occur and a robust set of clean energy technologies are developed, such restrictions will not inhibit a clean energy revolution during this century.

There are also environmental concerns associated with wind and solar energy. Wind power has noise and visual impacts that can limit its acceptance in some regions. Impact on bird life is also of concern. Some solar PV technologies contain hazardous materials (e.g., cadmium) that require safety standards in production and disposal/recycling. These concerns should be recognized and addressed, but they are relatively minor compared to the adverse environmental impacts associated with conventional energy technologies (Turkenburg 2000).

Transportation presents some unique and daunting challenges to the realization of a clean energy future. This is due to the rapid growth of personal vehicles and transportation energy use, and the very high dependence on petroleum-based transportation fuels in most (but not all) countries today.[3] As discussed in the next section, fuel cell vehicles and hydrogen fuel supply are two of the more promising long-term options for a clean energy revolution with respect to transportation.

The Transportation Challenge

The size of the world's passenger vehicle fleet increased 10-fold over the past 50 years, reaching 700 million vehicles by 1998. With rising population and income around the world, the number of passenger vehicles is projected to increase by a factor of 3 to 5 over the next 50 years (Birky et al. 2001). This implies dramatically higher transportation fuel use, in particular oil use, unless steps are taken to introduce much more efficient vehicles and new types of fuels.

Vehicle use is increasing especially rapidly in developing countries due to their very small fleet of motorized vehicles at the present time. China, for example, has about one motorized vehicle per 100 persons, less than the level reached in the United States in 1912 (Birky et al. 2001). Transportation sector energy use and CO_2 emissions are projected to increase by 140 percent in developing countries between 1997 and 2020, much faster than the projected rates of growth in the OECD or transition economies (IEA 2000a).

Previous chapters featured some policies for reducing oil use and pollutant emissions in transportation, including the ethanol fuel program in Brazil, vehicle efficiency standards in the United States, and the voluntary agreement to reduce new vehicle energy use and CO_2 emissions in the European Union. Also, some examples of effective urban transportation and land use planning were highlighted. But in general there has been less adoption of policies to stimulate cleaner transportation fuels and increased efficiency in transportation, compared to efforts made to promote renewable sources of electricity or increase energy efficiency in industry or buildings.

Auto manufacturers have developed a number of promising motor vehicle technologies, including higher-efficiency conventional vehicles, hybrid electric vehicles that contain a small internal combustion engine as well as an electric energy storage system, and fuel cell vehicles. Hybrid electric vehicles with very low pollutant emissions and efficiencies of up to twice those of conventional vehicles are starting to be mass produced. Fuel cell vehicles will be commercialized within a few years (Friedman et al. 2001). A combination of expanded RD&D, financial incentives, voluntary agreements with manufacturers, and mandatory efficiency and emissions standards should be pursued throughout the world so that markets for these innovative vehicles grow rapidly, costs are reduced through economies of scale and learning effects, and the overall fuel economy of new vehicles increases by 50 percent or more over the next decade and more over the long run.

Increasing the efficiency of new motor vehicles presents an excellent opportunity for international policy coordination and joint action since cars are increasingly produced and marketed on a global scale by multinational companies. Unfortunately, no serious effort along these lines is under way or under discussion outside of the European Union, due in large part to the lack of a global institution or mechanism for facilitating this type of action. This shortcoming could be overcome if an International Energy Efficiency and Renewable Energy Agency is created. In addition, policies for increasing the efficiency of new motor vehicles could be combined with policies for reducing the tailpipe emissions that harm public health (EF 2001).

Increasing vehicle efficiency by itself will not solve our transport energy problems because of the expected growth in the world's vehicle fleet and the heavy reliance on petroleum products today. New transport fuels, including greater use of natural gas (or other fuels derived from natural gas), in the short to medium term as well as fuels derived from renewable energy sources are needed. Regarding the latter, leading candidates include ethanol or methanol produced from cellulosic biomass, and hydrogen produced from solar power, wind power, or other forms of renewable energy. It is not yet possible to determine which of these fuels is preferable as the primary substitute for gasoline and diesel fuel, taking into account both economic and environmental considerations (Ahman, Nilsson, and Johansson 2001). Therefore, vigorous RD&D on a variety of alternative fuels should continue. But RD&D should be complemented by market creation policies such as financial incentives for early producers or adopters, public sector investment in distribution infrastructure, and renewable fuel content standards so that renewable-based fuels can "break into" the vehicle fuels market relatively soon in many more countries.

Fuel cell vehicles could play a major role in addressing the world's transportation challenges if their cost can be sufficiently reduced.[4] Fuel cells are powered by hydrogen, which can be obtained from a variety of sources, including fossil fuels or renewable energy. Fuel cell vehicles that carry hydrogen onboard can be highly efficient while producing only water vapor from the tailpipe. Fuel cell vehicles can also include an onboard fuel reformer that produces hydrogen from natural gas, methanol, or even gasoline, meaning that it is not necessary to supply hydrogen and store it onboard. However, the reformer is costly and inefficient, and it produces a wide range of pollutants, especially if gasoline is used as the fuel (Dunn 2001).

Introducing dedicated hydrogen fuel cell vehicles is particularly desirable from an energy and environmental perspective, as opposed to pursuing fuel cell vehicles that run on gasoline or methanol via a fuel reformer (Jensen and Ross 2000). Initially hydrogen could be produced from natural gas on a decentralized scale using the existing natural gas distribution infrastructure. As the fleet of fuel cell vehicles grows and the cost of renewable energy declines, hydrogen could be produced from renewable energy sources such as biomass or renewable-based electricity. A substantial hydrogen distribution infrastructure would be needed if hydrogen is produced in biomass refineries. But hydrogen could be produced in a decentralized manner, possibly at fueling stations, if electricity is used.

A coordinated strategy is needed to introduce fuel cell vehicles and hydrogen fuel, and at the same time steer investments in fuel cell vehicles and related fuel technologies to the cleanest options. Financial incentives and infrastructure investment can be justified based on the long-term social and environmental benefits that would result from a hydrogen-based fuel cell vehicle revolution in both industrialized and developing countries. A market transformation strategy for hydrogen-based fuel cell vehicles might consist of higher taxes on gasoline and diesel fuel in countries where market prices do not fully account for environmental and social costs, financial incentives for production and purchase of innovative vehicles, bulk procurement to support initial markets, and mandatory energy efficiency or CO_2 emissions standards for new vehicles (Ogden, Williams, and Larson 2001).

Reducing growth in the number and use of cars and light trucks is another element of a comprehensive strategy for "greening" transportation. There are a wide range of options and policies for restraining motor vehicle use including:

- higher fuel and/or vehicle taxes,
- charging for vehicle insurance "at the pump" as an add-on to fuel purchases,
- expanding and improving the quality of public transportation,
- promoting telecommuting and e-commerce,
- providing safe and convenient options for bicyclists and walkers,
- restricting urban sprawl, and
- integrating transportation and land use planning.

It is estimated that these policies in combination could cut overall vehicle use in the United States, for example, by about 15 percent over the long run (Birky et al. 2001, Doniger et al. 2002). These policies can also result in significant reductions in future motor vehicle use in developing countries where transportation infrastructure and vehicle fleets are growing rapidly, thereby improving air quality, reducing congestion, and saving energy (Bose 1998, Reddy, Anand, and D'Sa 2000).

Combining vigorous vehicle efficiency improvements, a shift to cleaner fuels and eventually hydrogen-based fuel cell vehicles, and actions to limit growth in personal vehicle use can resolve the long-term transportation challenge, at least with respect to personal transport.[5] For example, an analysis prepared for the U.S. Department of Energy shows that aggressive fuel efficiency improvements coupled with introduction of both bioethanol and hydrogen-based fuel cell vehicles could cut total energy use of U.S. passenger vehicles in 2050 by 64 percent and CO_2 emissions by 96 percent relative to projected levels with a stagnant fuel economy and continued reliance on oil as the predominant vehicle fuel (Birky et al. 2001).

The transportation challenge is due in no small part to the type of vehicles chosen by individuals and the type of infrastructure chosen by society as a whole, in particular the car-oriented culture well-established in the United States and being copied for better or worse in other countries. Vehicle and transportation choices reflect, in part, income levels (e.g., many families purchase motorized vehicles when their income permits this). But personal choice also plays a role, especially in cities with convenient, high-quality public transportation systems. In these cities, individuals have different transportation options. Personal choice can also influence future energy demand through decisions about family size and lifestyle.

Population Growth and Lifestyle Choice

Many social and cultural developments will affect future energy resources and consumption levels, and consequently the difficulty of achieving a clean energy revolution during this century. Population growth and lifestyle choice are two important factors in this regard. Although these factors are normally taken as a "given" in energy policy analysis, policies to limit population growth worldwide and encourage less materially consumptive lifestyles in already affluent regions could help to facilitate a clean energy revolution.

The world's population is expected to increase about 50 percent by 2050 under "medium growth" assumptions, with almost all the increase expected in developing countries (Box 8-2). But population growth could be held to 35 to 40 percent during this period if efforts to lower fertility are stepped up. Reducing population growth will facilitate a transition to a sustainable future by reducing consumption of energy and other resources. At the same time, providing modern energy sources to households now lacking electricity and modern fuels will improve standards of living and help to reduce population growth in developing countries (Reddy 2000). Other policies that can reduce population growth

BOX 8-2

Limiting Population Growth

The world population doubled since 1960 to 6.1 billion as of 2001. The global effort to limit climate change to acceptable levels and improve standards of living in developing countries will be profoundly influenced by population growth. While the rate of increase has declined significantly in recent years, the world's population is still rising at a rate of 1.2 percent per year. The United Nations projects that the world population will reach 9.3 billion by 2050 in its medium growth projection (UNFPA 2001). Nearly all of this growth is expected to occur in developing countries, which by 2050 are expected to contain over 85 percent of the world's population.

It is estimated that about half of the growth in world energy demand between 1890 and 1990 was due to population growth and the other half was due to increasing energy use per capita (Reddy 2000). In the future, limiting population growth will reduce global energy consumption. And improving the quality of energy supplies and efficiency of energy use in developing countries, especially in rural areas, will increase standards of living and thereby help to lower fertility rates and population growth. Thus the "energy-population nexus" has multiple and interrelated dimensions (Reddy 2000).

Nations such as Bangladesh, South Korea, Singapore, Thailand, and Colombia have experienced rapid declines in fertility rates through implementing policies to reduce population growth, including better health care and education of women and expanded family planning services (WRI 1998). But 350 million women worldwide still lack access to any contraceptive services, and consequently nearly 40 percent of pregnancies worldwide are unwanted, at least by women (Engelman, Halweil, and Nierenberg 2002). Health care and family planning initiatives are relatively modest in cost and provide broad benefits including poverty alleviation, less environmental degradation, and reduced emissions of CO_2 and other gases causing global warming (UNFPA 2001).

Reducing the population of wealthier nations is also important, given the relatively high per capita consumption of energy and other resources in these countries. In the United States, for example, the roughly 50 million people expected to be added over the next 20 years (due mainly to immigration) will use more energy than is consumed in either Africa or India at their current consumption rates. Population growth in Western nations can be reduced by providing better education and employment opportunities to low-income families, as well as by expanding family planning services to all citizens. Also, education campaigns can help to convince couples to have only one or two children (Brown, Gardner, and Halweil 1998).

How much difference could efforts to limit population growth make? A U.S. National Academy of Sciences study concluded that a 10 percent reduction in world population by 2050, compared to the level anticipated under business-as-usual policies and trends, is realistic and attainable (NAS 1999). This would mean nearly a billion fewer people consuming energy and other natural resources, thereby facilitating a transition to a sustainable energy future. However, the world would still need to increase energy efficiency and shift to cleaner energy sources on a massive scale to address the problems outlined in Chapter 1.

◉ ◉ ◉

include better health care and education especially of women, expanding family planning services, and efforts to convince couples to limit the number of children they have (Kates 2000, UNFPA 2001).

Lifestyle affects energy consumption through many choices such as the types of appliances, homes, and vehicles families purchase, how they are used, and where people choose to live. In the United States, new homes are increasing in size, energy-consuming devices are proliferating, personal vehicles are getting larger and more powerful, and urban sprawl is continuing in many cities. This has caused the "growth effect" to outstrip the "efficiency effect," leading to rising energy consumption and CO_2 emissions in the United States in spite of high rates of energy efficiency improvement in recent years.

Personal choices can have a significant effect on the total consumption of energy and the rate of energy demand growth (Lutzenhiser and Grossard 2000, Schipper 1991). Combating the desire for more consumption of material goods and a "bigger is better" mentality in already affluent nations could play a role in a clean energy revolution during the twenty-first century. Less materially oriented lifestyles could also lead to higher levels of personal fulfillment by reducing the amount of work necessary, thereby providing more leisure time, and by placing greater emphasis on other aspects of life such as human relations, community involvement, creativity, exercise and interaction with nature, and spirituality (Raskin et al. 2002).

Changing lifestyle choices on a large scale in affluent nations is not simple or easy. Many institutions ranging from business interests, advertising, and popular culture (TV, films, etc.) promote materially oriented, "bigger is better" lifestyles. In the United States, a sizable fraction of the public espouses concern for the environment and disaffection with materialism (Ray and Anderson 2000). But relatively few people have modified their behavior or lifestyle in ways that are consistent with these views. Educational efforts, use of role models, as well as more conventional policies such as taxes and financial incentives can influence behavior and lifestyle choices (Gardner 2001). At the same time, more research is needed concerning the most effective ways to change values and promote lifestyle choices that are less materially consumptive and more consistent with a sustainable future (Lutzenhiser and Grossard 2000).

Conclusion

Past global energy transitions occurred over many decades, driven in large part by technological advances. For example, coal production expanded for over a century before it overtook wood as the world's leading energy source in the late 1800s. Likewise, petroleum production expanded for nearly a century before it overtook coal as the world's leading source of energy during the 1960s. An important characteristic of these past energy transitions is that resource scarcity played only a minor role. New energy sources overtook older sources because they were more desirable due in part to the emergence of innovative conversion and end-use technologies—coal as high-energy-density fuel burned in steam engines and turbines, and oil as a fuel for automobiles, aircraft, and other vehicles (Grubler 1998).

Likewise the transition to modern renewable energy resources during the twenty-first century will take many decades if not a full century to complete, assuming a worldwide commitment is made and comprehensive and strong policies are adopted to foster this transition. Like past transitions, the pace will be driven to a large degree by the characteristics of energy conversion and end-use technologies. Some of these technologies, such as wind turbines, ethanol fuel production, highly efficient appliances, and highly efficient light sources, are already widely produced (although still evolving). Some, such as PV electric systems and hybrid vehicles, are commercially available but produced on a relatively limited scale. Other technologies, such as fuel cell vehicles, biorefineries, and hydrogen fuel technologies, are still under development, while others remain to be discovered. These technologies will overtake fossil fuels and inefficient end-use devices because they have superior attributes and the barriers inhibiting their adoption are overcome, not because the world will run out of oil, coal, gas-guzzling vehicles, inefficient appliances, or incandescent lamps.

The clean energy transition will take many decades, in part because of the long lives and slow turnover of some key energy-supply and end-use technologies. New homes and commercial buildings last 50 to 100 years; new electric power plants last 30 to 50 years (PCAST 1997). Constructing inefficient buildings or coal-fired power plants today will influence energy use for many decades. Even if all new power sources were renewable-based, coal will remain as an electricity source and boiler fuel for many decades, especially in countries such as China that are now heavily dependent on coal. Likewise, it will take a number of decades for innovative vehicle technologies such as fuel cell vehicles to dominate the entire vehicle stock (Ogden, Williams, and Larson 2001). In addition, urban design today will influence energy consumption for centuries (IEA 2001h).

These time constraints, sometimes referred to as energy system inertia, suggest that it is critical to hasten rather than delay a clean energy transition. The transition should start now, as portrayed in the Global Clean Energy Scenario, to prevent "lock-in" of inefficient and highly polluting infrastructure and technology, especially in rapidly growing developing nations. Delaying the transition would increase pollutant emissions and increase the risk of catastrophic climate change (O'Neill and Oppenheimer 2002). Delaying would also limit the technological advances that result from "learning by doing," thereby making a transition down the road more difficult and costly (Schneider and Azar 2001).

Accelerating the transition—in effect, sparking a clean energy revolution—would provide another important benefit. It would build "coalitions for change." Such coalitions are needed to overcome the political obstacles preventing the adoption of comprehensive and strong policies in some locations—most notably in the United States (Gardner 2001). As explained in Chapter 5, the U.S. automotive, utility, and fossil fuel industries have been able to block the adoption of key national policies for advancing energy efficiency and renewable energy use. But moving ahead with the transition in more progressive countries, states, and municipalities will build up clean energy industries as well as the understanding of and support for these policies and technologies among civil society. As clean energy coalitions grow in economic and political influence, eventually they should prevail in difficult political battles such as the fight for strong vehicle efficiency standards, renewable energy obligations, and commitments to cut greenhouse gas emissions at the national level in the United States.[6]

A clean energy revolution cannot be accomplished unilaterally or through action by one block of nations. All countries, rich and poor, must cooperate sooner or later. Different countries may have differing motives, pursue different policies, and use different sets of technologies, but all (or nearly all) countries would benefit. In a world where armed conflicts, resource conflicts, and social conflicts threaten the security of all citizens, nations and individuals uniting behind a common goal and working together to achieve this goal could be an important indirect benefit of a global clean energy revolution.

Daniel Yergin has called the twentieth century "the century of oil" (Yergin 1991). In fact it was really the century of oil and coal. How will the twenty-first century be known? It could be the century of renewable energy sources complemented by natural gas as the world makes a transition away from fossil fuels—a century of healthier air, moderate global warming, lower cost of energy services, greater security and reliance on local energy resources, and improved standards of living for all citizens of the world. Or it could be a century of continued heavy dependence on fossil fuels including oil and coal—a century of worsening air quality, accelerating and potentially catastrophic global warming, greater insecurity as oil resources are depleted, higher cost of energy services, and continued impoverishment for a large number of the world's citizens. The policies and strategies for achieving a more sustainable energy future are at hand. The choice of an energy future is ours to make.

Notes

1. About 50 GW of PV capacity would be required to provide all of these households with 100 watt PV systems.
2. These estimates were provided by the Tellus Institute using the Model for the Assessment of Greenhouse-Gas Induced Climate Change. In addition to CO_2 emissions, it is assumed that the radiative forcing from all other sources is equal to about $1W/m^2$ in 2100 (Schneider and Azar 2001).
3. Petroleum provides about 97 percent of the energy consumed by the transportation sector in the United States, for example. Brazil is exceptional in that petroleum only provides 85 percent of overall transportation energy consumption.
4. The issues related to the development, introduction, and possible evolution of fuel cell vehicles and hydrogen fuel are quite complex. A detailed examination of these issues is beyond the scope of this book, but interested readers should refer to Dunn 2001, Lovins and Williams 1999, NHA 2001, and Ogden, Williams, and Larson 2001.
5. Similar strategies and policies to increase vehicle efficiency, introduce cleaner fuels, and stimulate modal shifts are also needed for freight transport (Birky et al. 2001).
6. Reforming the political process (e.g., by limiting the influence of large corporations through campaign finance reform, increasing the openness of decision making, and expanding direct citizen involvement in setting policy), should also facilitate the adoption of policies that will accelerate the clean energy revolution.

Appendix

Key Assumptions in the Global Clean Energy Scenario

- Economic growth over the course of the century averages 2.2 percent per annum worldwide—1.7 percent per annum in Organization for Economic Cooperation and Development (OECD) countries, 2.5 percent per annum in the transition economies, and 3.2 percent per annum in developing countries. These are roughly the same growth rates in the International Institute for Applied Systems Analysis–World Energy Council (IIASA–WEC) ecologically driven scenario, which assumes that priority is given to more equitable distribution of wealth worldwide.
- Energy intensity (energy use per unit of gross domestic product) declines 2.2 percent per annum in OECD countries, 2 percent in transition countries, and 2 percent in developing countries. The OECD rate is more rapid than the 1.5 percent per annum decline that occurred between 1973 and 2000, but it has been achieved at times in the past and is feasible if concerted efforts to improve energy efficiency are made (see Chapter 5, for example). The transition and developing country rates are well below what China has achieved for decades. These rates of improvement are feasible if concerted modernization and energy efficiency efforts are maintained over the long run.
- Renewable energy supply expands 3 percent per annum in OECD and developing countries and 4 percent per annum in transition countries through 2050. The rate is higher for transition countries because of

their low starting point. Developing countries both increase their renewable energy use and shift from traditional to modern forms of renewable energy. After 2050, renewable energy growth moderates to 2 percent per annum in OECD and developing countries, and 3 percent per annum in transition countries. These growth rates are consistent with the country- and technology-specific growth rates mentioned in the main text (i.e., wind and solar power growing 20–30 percent per year for at least two decades). In fact, renewable energy supply could expand as much as 4 percent annually in the United States over the next 20 years if strong national policies in support of renewables were enacted (see Chapter 5).

- Natural gas use increases 2 percent per annum through 2020, remains constant from 2020 to 2060, and gradually declines after 2060. The rate of growth in natural gas use in the near term is not excessive relative to projected growth rates in business-as-usual energy forecasts. These forecasts project worldwide natural gas use increasing 3 to 4 percent annually through 2020 (IEA 2000a, EIA 2001b).
- Consumption of oil, coal, and nuclear energy declines in absolute terms throughout the century, consistent with these other assumptions. It is assumed that coal use drops 32 percent by 2030 and 75 percent by 2050. With the growth of renewable energy and natural gas as electricity sources, coal use could decline even faster (see Chapter 5, for example). Oil use is more problematic, but declines about 1 percent per annum over the next 70 years assuming strong efforts are made to increase vehicle efficiency, introduce cleaner fuels, and limit growth in personal motorized vehicle use. Oil consumption is assumed to decline more rapidly after 2070 as renewable-based fuels fully mature. Nuclear energy is phased out over the next 50 years as existing nuclear plants are retired.

References

ACEA [Association des Constructeurs Europeens d'Automobiles]. 1998. ACEA Commitment on CO_2 Emission Reductions from New Passenger Cars in the Framework of an Environmental Agreement between the European Commission and ACEA. Brussels, Belgium: Association des Constructeurs Europeens d'Automobiles.

———. 2002. European Automotive Industry Further Reduces New Car CO_2 Emissions in 2001. Brussels, Belgium: Association des Constructeurs Europeens d'Automobiles. www.acea.be/acea/20020709PressRelease.pdf.

Agarwal, A. 1999. *Making the Kyoto Protocol Work.* New Delhi, India: Centre for Science and Environment.

Ahman, M., L. J. Nilsson, and B. Johansson. 2001. Cars and Fuels of Tomorrow: Strategic Choices and Policy Needs. In *Proceedings of the 2001 ECEEE Summer Study* 1: 528–38. Paris: European Council for an Energy-Efficient Economy.

AID [Agency for International Development]. 1996. *Strategies for Financing Energy Efficiency.* Washington, DC: U.S. Agency for International Development. July.

Aitken, D. 1998. *Putting It Together: Whole Buildings and a Whole Buildings Policy.* Washington, DC: Renewable Energy Policy Project.

ALA [American Lung Association]. 2001. *Trends in Air Quality.* New York: American Lung Association. August.

Almeida, M. A., R. Schaeffer, and E. L. la Rovere. 2001. The Potential for Electricity Conservation and Peak Load Reduction in the Residential Sector of Brazil. *Energy—The International Journal 26* (4): 413–429.

Aranda, F.A. and I.C. Cruz. 2000. Breezing Ahead: The Spanish Wind Energy Market. *Renewable Energy World* 3 (3): 35–45.

AWEA [American Wind Energy Association]. 2002. *Global Wind Energy Market Report.* Washington, DC: American Wind Energy Association.

Azuaga, D. 2000. Danos ambientais causados por veículos leves no Brasil. Master's thesis, Programa de Planjemento Energetico, COPPE/UFRJ. (Rio de Janeiro: Universidade Federal do Rio de Janeiro).

Bailie, A., S. Bernow, W. Dougherty, M. Lazarus, and S. Kartha. 2001. *The American Way to the Kyoto Protocol: An Economic Analysis to Reduce Carbon Pollution.* Washington, DC: World Wildlife Fund.

Bakthavatsalam, V. 2001. Windows of Opportunity: IREDA and the Role of Renewable Energy in India. *Refocus* May, pp. 12–15.

Balu, V. 1997. Issues and Challenges Concerning Privatisation and Regulation in the Power Sector. *Energy for Sustainable Development* 3 (6): 6–13.

Banerjee, N. 2001. Fears, Again, of Oil Supplies at Risk. *New York Times* Oct. 14, section 3, pp. 1, 11.

Bang, K. 2000. ESCO Market in Korea. Presentation at the National Association of Energy Service Companies Conference, Houston, TX. May 17, 2000.

[BCAP] Building Codes Assistance Project. 2001. Residential Building Code Status and Commercial Building Code Status. Washington, DC: Building Codes Assistance Project. www.bcap-energy.org.

Beck, P.W. 2001. Nuclear Energy in the Twenty-First Century: Examination of a Contentious Subject. *Annual Review of Energy and Environment* 24: 113–138.

Bentley, R.W. 2002. Global Oil and Gas Depletion: An Overview. *Energy Policy* 30: 189–205.

Bernow, S., K. Cory, W. Dougherty, M. Duckworth, S. Kartha, and M. Ruth. 1999. *America's Global Warming Solutions.* Washington, DC: World Wildlife Fund.

Bernow, S., M. Fulmer, I. Peters, M. Ruth, and D. Smith. 1997. *Ecological Tax Reform: Carbon Taxes with Tax Reductions in New York.* Boston: Tellus Institute.

Bernow, S., S. Kartha, M. Lazarus, and S. Page. 2000. *Cleaner Generation, Free Riders, and Environmental Integrity: Clean Development Mechanism and the Power Sector.* Boston: Tellus Institute.

Bernstein, M., R. Lempert, D. Loughran, and D. Ortiz. 2000. *The Public Benefit of California's Investments in Energy Efficiency.* MR-1212.0-CEC. Santa Monica, CA: Rand Corporation.

Berry, T. and M. Jaccard. 2001. The renewable portfolio standard: design considerations and an implementation survey. *Energy Policy* 29: 263–277.

Bertoldi, P., P. Waide, and B. Lebot. 2001. Assessing the Market Transformation for Domestic Appliances Resulting from European Union Policies. In

Proceedings of the 2001 ECEEE Summer Study 2: 191–202. Paris: European Council for an Energy-Efficient Economy.

Birky, A., D. Greene, T. Gross, D. Hamilton, K. Heitner, L. Johnson, J. Maples, J. Moore, P. Patterson, S. Plotkin, and F. Stodolsky. 2001. *Future U.S. Highway Energy Use: A Fifty Year Perspective.* Washington, DC: U.S. Department of Energy, Office of Transportation Technologies.

Birner, S. 2000. How Thailand Washed Away Wasteful Lighting. *IAEEL Newsletter* 24(9). Stockholm, Sweden: International Association for Energy-Efficient Lighting.

Birner, S. and E. Martinot. 2002. *The GEF Energy-Efficient Product Portfolio: Emerging Experience and Lessons.* Washington, DC: Global Environmental Facility.

Blackman, A. and X. Wu. 1999. Foreign Direct Investment in China's Power Sector: Trends, Benefits, and Barriers. *Energy Policy* 27: 695–711.

Bluestein, J. and M. Lihn. 1999. Historical Impacts and Future Trends in Industrial Cogeneration. In *Proceedings of the 1999 ACEEE Summer Study on Energy Efficiency in Industry* 479–501. Washington, DC: American Council for an Energy-Efficient Economy.

Borg, N. 2002. Personal communication from Nils Borg, International Association for Energy Efficient Lighting, Stockholm, Sweden.

Bose, R. K. 1998. Automotive Energy Use and Emissions Control: A Simulation Model to Analyse Transport Strategies for Indian Metropolises. *Energy Policy* 26: 1001–1016.

[BP] British Petroleum. 2001. *BP Statistical Review of World Energy 2001.* London: British Petroleum.

Bradsher, K. 2000. General Motors Raises Stakes in Fuel Economy War with Ford. *New York Times.* August 3.

Brown, L. R., G. Gardner, and B. Halweil. 1998. *Beyond Malthus: Sixteen Dimensions of the Population Problem.* Washington, DC: Worldwatch Institute.

Brown, M. A. 2001. Market Failures and Barriers as a Basis for Clean Energy Policies. *Energy Policy* 29: 1197–1208.

BTM Consult 2001. A Towering Performance: Latest BTM Report on the Wind Industry. *Renewable Energy World* 4 (4): 68–87.

Calwell, C., J. Zugel, P. Banwell, and W. Reed. 2002. 2001—A CFL Odyssey: What Went Right. *Proceedings of the 2002 ACEEE Summer Study on Energy Efficiency in Buildings* 6: 15–27. Washington, DC: American Council for an Energy-Efficient Economy.

Cameron, M., J. Stierstorfer, and D. Chiaramonti. 1999. Financing Renewable Energy. *Renewable Energy World* 2 (4): 41–45.

Cameron, M., J. Stierstorfer, S. Teske, and C. Aubrey. 2001. Solar Generation: A Blueprint for Growing the PV Market. *Renewable Energy World* 4 (5).

Cameron, M., J. M. Wilder, and M. Pugliese. 2001. From Kyoto to Bonn: Implications and Opportunities for Renewable Energy. *Renewable Energy World* 4 (5).

Campbell, C. J. and J. H. Laherrere. 1998. The End of Cheap Oil. *Scientific American* 278 (3): 78–83.

CARB [California Air Resources Board]. 2001. ARB Holds to ZEV Mandate. News Release. Sacramento, CA: California Air Resources Board. www.arb.ca.gov/newsrel/nr012601.htm.

Carvalho, I. M. 1997. Greenhouse Gas Emissions and Bio-Ethanol Production/Utilization in Brazil. In *South–South North Partnership on Climate Change and Greenhouse Gas Emissions.* Edited by S. K. Ribeiro and L. P. Rosa. Rio de Janeiro: COPPE, Federal University of Rio de Janeiro.

———. 1998. The Role of Copersucar in Improving Technology for Ethanol Production from Sugar Cane in Sao Paulo. Paper presented at the STAP Workshop on Technology Transfer in the Energy Sector, Amsterdam, Jan. 19–20.

Casten, T. R. 1998. Turning Off the Heat: *Why America Must Double Energy Efficiency To Save Money and Reduce Global Warming.* Amherst, NY: Prometheus Books.

Casten, T. R. and M. C. Hall. 1998. *Barriers to Deploying More Efficient Electrical Generation and Combined Heat and Power Plants.* White Plains, NY: Trigen Energy Corp.

Cavanagh, R. 1999. Congress and Electric Industry Restructuring: Environmental Imperatives. *Electricity Journal* 12 (7): 11–20.

———. 2001. Revisiting the "Genius of the Marketplace": Cures for the Western Electricity and Natural Gas Crises. *Electricity Journal* 14 (5): 11–18.

CEC [California Energy Commission]. 1999. *The Energy Efficiency Public Goods Charge Report: A Proposal for a Millennium.* Sacramento: California Energy Commission.

———. 2001a. *Costs and Benefits of a Biomass-to-Ethanol Production Industry in California.* P500-01-002. Sacramento: California Energy Commission.

———. 2001b. *Investing in Renewable Electricity Generation in California.* P500-00-022. Sacramento: California Energy Commission.

———. 2002. *The Summer 2001 Conservation Report.* Sacramento: California Energy Commission.

Cecelski, E. W. 1995. From Rio to Beijing: Engendering the Energy Debate. *Energy Policy* 23: 561–575.

Center for Responsive Politics. 2001. Web site of the Center for Responsive Politics, Washington, DC. www.opensecrets.org.

Chandler, W., J. Parker, I. Bashmakov, Z. Genshev, J. Marousek, S. Pasierb, M. Raptsun, and Z. Dadi. 1999. *Energy Efficiency Centers in Six Countries: A Review.* PNNL-13073. Washington, DC: Pacific Northwest National Laboratory, Advanced International Studies Unit.

Clark, A. 2000. Demand-Side Management Investments in South Africa: Barriers and Possible Solutions for New Power Sector Contexts. *Energy for Sustainable Development* 4 (4): 27–35.

———. 2001. Making Provision for Energy-Efficiency Investment in Changing Markets: an International Review. *Energy for Sustainable Development* 5 (2): 26–38.

Claussen, E. 2001. Climate Change: A Strategy for the Future. Speech by Eileen Claussen at the University of Rhode Island, Sept 25. Arlington, VA: PEW Center on Global Climate Change.

Claussen, E. 2002. Climate Change: Myths and Realities. Speech by Eileen Claussen at the "Climate in North America" Conference, New York, NY, July 17. Arlington, VA: PEW Center on Global Climate Change.

Clean Air Task Force 2000. *Death, Disease, and Dirty Power: Mortality and Health Damage Due to Air Pollution from Power Plants.* Boston, MA: Clean Air Task Force. October.

Clemmer, S., D. Donovan, A. Nogee, and J. Deyette. 2001. *Clean Energy Blueprint: A Smarter National Energy Policy for Today and the Future.* Cambridge, MA: Union of Concerned Scientists.

COPPE. 1998. *Estimativa do Potencial de Conservacao de Energia Eletrica pelo Lado de Demanda no Brasil.* Rio de Janeiro, Brazil: Universidade Federal do Rio de Janeiro, Instituto Alberto Luiz Coimbra de Pos-Graduacao e Pesquisa de Engenharia, Programa de Planejamento Energetico.

Cowart, R. 2001. *Efficient Reliability: The Critical Role of Demand-Side Management Resources in Power Systems Markets.* Montpelier, VT: Regulatory Assistance Project.

Crowley, S. 2001. Partnerships Leading the Way Forward to Ensure Quality Energy Efficiency Installers. In *Proceedings of the 2001 ECEEE Summer Study* 2: 278–285. Paris: European Council for an Energy-Efficient Economy.

Dahl, C. and K. Kuralbayeva. 2001. Energy and the Environment in Kazakhstan. *Energy Policy* 29: 429–440.

Davis, S. 2001. *Transportation Energy Data Book: Edition 21.* ORNL-6966. Oak Ridge, TN: Oak Ridge National Laboratory.

de Beer, J., E. Worrell, and K. Blok. 1998. Future Technologies for Energy-Efficient Iron and Steel Making. *Annual Review of Energy and Environment* 23:123–206.

DeCanio, S. 1993. Barriers within Firms to Energy-Efficient Investments. *Energy Policy* 21: 906–14.

DeCicco, J., F. An, and M. Ross. 2001. *Technical Options for Improving the Fuel Economy of U.S. Cars and Light Trucks by 2010–2015.* Washington, DC: American Council for an Energy-Efficient Economy.

DeCicco, J., R. Diamond, S. L. Nolden, J. DeBarros, and T. Wilson. 1995. *Improving Energy Efficiency in Apartment Buildings.* Washington, DC: American Council for an Energy-Efficient Economy.

Deffeyes, K. S. 2001. *Hubbert's Peak: The Impending World Oil Shortage.* Princeton, NJ: Princeton University Press.

den Elzen, M. G. J. and A. P. G. de Moor. 2001. *Evaluating the Bonn Agreement and Some Key Issues.* Bilthoven, Netherlands: RIVM National Institute of Public Health and the Environment.

Dodds, D., E. Baxter, and S. Nadel. 2000. Retrocommissioning Programs: Current Efforts and Next Steps. In *Proceedings of the 2000 ACEEE Summer Study on Energy Efficiency in Buildings* 4: 79–93. Washington, DC: American Council for an Energy-Efficient Economy.

DOE [U.S. Department of Energy]. 2000. *Clean Energy Partnerships: A Decade of Success.* DOE/EE-0213. Washington, DC: U.S. Department of Energy.

———. 2001. *Department of Energy Historic Budget Authority by Organization.* Washington, DC: U.S. Department of Energy.

Doniger, D., D. Friedman, R. Hwang, D. Lashof, and J. Mark. 2002. *Dangerous Addiction: Ending America's Oil Dependence.* Washington, DC: Natural Resources Defense Council and Union of Concerned Scientists.

Dooley, J. J. 1998. Unintended Consequences: Energy R&D in a Deregulated Energy Market. *Energy Policy* 26 (7): 547–555.

DOS [U.S. Department of State]. 2002. U.S. Climate Action Report 2002. Washington, DC: U.S. Department of State.

DSIRE. 2001. *Database of State Incentives for Renewable Energy.* www.dsireusa.org.

Duke, R. D., A. Jacobson, and D. M. Kammen. 2002. Photovoltaic Module Quality in the Kenyan Solar Home Systems Market. *Energy Policy* 30: 477–500.

Dunn, S. 2001. *Hydrogen Futures: Toward a Sustainable Energy System.* Worldwatch Paper 157. Washington, DC: Worldwatch Institute.

Dunn, S. and C. Flavin. 2002. Moving the Climate Change Agenda Forward. In *State of the World 2002,* 24–50. Edited by L. Starke. New York: W. W. Norton.

Dutt, G. S., F. G. Nicchi, and M. Brugnoni. 1997. Power Sector Reforms in Argentina: An Update. *Energy for Sustainable Development* 3 (6): 36–54.

Dutta, S., I. H. Rehman, P. Malhotra, and V. Ramana. 1997. *Biogas: The Indian NGO Experience.* New Delhi: Tata Energy Research Institute.

DWIA [Danish Wind Industry Association]. 2002. *Danish Wind Power 2001.* Copenhagen: Danish Wind Industry Association. www.windpower.org/news/stat2001.htm.

Edjekumhene, I., M. B. Amadu, and A. Brew-Hammond. 2001. Preserving and Enhancing Public Benefits under Power Sector Reform: The Case of Ghana. *Energy for Sustainable Development* 5 (2): 39–47.

EERE. [Office of Energy Efficiency and Renewable Energy]. 2000. *Clean Energy Partnerships: A Decade of Success.* Washington, DC: U.S. Department of Energy, Office of Energy Efficiency and Renewable Energy.

EF [Energy Foundation]. 2001. Bellagio Memorandum on Motor Vehicle Policy: Principles for Vehicles and Fuels in Response to Global Environmental and Health Imperatives. San Francisco: Energy Foundation.

Egan, K. and P. du Pont. 1998. *Asia's New Standard for Success: Energy Efficiency Standards and Labeling Programs in 12 Asian Countries.* Washington, DC: International Institute for Energy Conservation.

Egger, C. and G. Dell. 1999. The Regional Energy Plan of Upper Austria: 12% CO_2 Reduction in 4 Years. In *Energy Efficiency and CO_2 Reduction: The Dimensions of the Social Challenge—1999 Summer Study Proceedings.* Paris: European Council for an Energy-Efficient Economy.

EIA [Energy Information Administration]. 1998. *Renewable Energy Annual 1998.* DOE/EIA-0603(98). Washington, DC: Energy Information Administration.

———. 1999. *Renewable Energy Issues and Trends 1998.* DOE/EIA-0628(98). Washington, DC: Energy Information Administration.

———. 2000a. *Long-Term World Oil Supply: A Resource Base/Production Path Analysis.* Washington, DC: Energy Information Administration, U.S. Department of Energy. www.eia.doe.gov/pub/oil_gas/...ons/2000/long_term_supply/index.htm.

———. 2000b. *Emissions of Greenhouse Gases in the United States 1999.* DOE/EIA-0573(99). Washington, DC: Energy Information Administration, U.S. Department of Energy.

———. 2000c. *Annual Energy Outlook 2001.* DOE/EIA-0383(2001). Washington, DC: U.S. Department of Energy, Energy Information Administration.

———. 2000d. *Electric Utility Demand Side Management 1999.* www.eia/gov/cneaf/electricity/dsm99. Washington, DC: U.S. Department of Energy, Energy Information Administration.

———. 2000e. *The National Energy Modeling System: An Overview 2000.* DOE/EIA-0581(2000). Washington, DC: U.S. Department of Energy, Energy Information Administration.

———. 2001a. *Annual Energy Review: 2000.* DOE/EIA-0384(2000). Washington, DC: U.S. Department of Energy, Energy Information Administration.

———. 2001b. *International Energy Outlook 2001.* DOE/EIA-0484(2001). Washington, DC: U.S. Department of Energy, Energy Information Administration.

———. 2001c. *Emissions of Greenhouse Gases in the United States 2000.* DOE/EIA-0573(2000). Washington, DC: U.S. Department of Energy, Energy Information Administration.

Eikeland, P. O. 1998. Electricity Market Liberalisation and Environmental Performance: Norway and the U.K. *Energy Policy* 26: 917–928.

Eletrobrás 1999. *Estimativa do Potencial de Cogeracao no Brasil.* Rio de Janeiro: Eletrobrás, Coordinating Group for Planning of Electricity Systems.

———. 2000. *Ten-Year Expansion Plan 2000/2009.* Rio de Janeiro: Eletrobrás, Engineering Directorate.

[ELI] Environmental Law Institute. 2000. *Cleaner Power: The Benefits and Costs of Moving from Coal to Natural Gas Power.* Washington, DC: Environmental Law Institute.

Engelman, R., B. Halweil, and D. Nierenberg. 2002. Rethinking Population, Improving Lives. In *State of the World 2002,* 127–148. Edited by L. Starke. New York: W. W. Norton.

ENS. 2001a. Climate Change Costs Could Top $300 Billion Annually. Environmental News Service. February 7. http://ens.lycos.com/ens/feb2001/2001L-02-05-02.html.

———. 2001b. Bullet Holes Spill Alaskan Oil. Environmental News Service. October 8. http://ens.lycos.com/ens/oct2001/2001L-10-08-06.html.

———. 2001c. 2001 the Second Warmest Year on Record. Environmental News Service. December 18. http://ens.lycos.com/ens/dec2001/2001L-12-18-01.html.

———. 2001d. Threat of Nuclear Terrorism is Growing, Experts Warn. Environmental News Service. November 2. http://ens.lycos.com/ens/nov2001/2001L-11-02-06.html.

EPA [U.S. Environmental Protection Agency]. 2000. *Light-Duty Automotive Technology and Fuel Economy Trends 1975 Through 2000.* EPA420-R00-008. Washington, DC: U.S. Environmental Protection Agency, Office of Transportation and Air Quality.

———. 2001. *The Power of Partnerships: Climate Protection Partnerships Division 2000 Annual Report.* EPA 430-R-01-009. Washington, DC: U.S. Environmental Protection Agency, Office of Air and Radiation.

———. 2002a. *National-Scale Air Toxics Assessment.* Washington, DC: U.S. Environmental Protection Agency.

———. 2002b. Inventory of U.S. Greenhouse Gas Emissions and Sinks:

1990–2000. Washington, DC: U.S. Environmental Protection Agency, Office of Atmospheric Programs.

Espey, S. 2001. Renewables Portfolio Standard: A Means for Trade with Electricity from Renewable Energy Sources? *Energy Policy* 29: 557–566.

Eto, J., C. Goldman, and S. Nadel. 1998. *Ratepayer-Funded Energy Efficiency Programs in a Restructured Electricity Industry: Issues and Options for Regulators and Legislators.* Washington, DC: American Council for an Energy-Efficient Economy.

Eto, J., R. Prahl, and J. Schlegel. 1996. *A Scoping Study on Energy-Efficiency Market Transformation by California Energy-Efficiency Programs.* LBNL-39058. Berkeley, CA: Lawrence Berkeley National Laboratory.

European Commission. 2000. *Directive of the European Parliament and of the Council on the Promotion of Electricity from Renewable Energy Sources in the Internal Electricity Market.* COM(2000) 884 final. Brussels: Commission of the European Communities.

Eurosolar. 2001. *Memorandum for the Establishment of an International Renewable Energies Agency (IRENA).* Bonn, Germany: European Association for Renewable Energies. Feb.

EWEA [European Wind Energy Association], Forum for Energy and Development, and Greenpeace International. 1999. *Wind Force 10: A Blueprint to Achieve 10% of the World's Electricity from Wind Power by 2020.* London: European Wind Energy Association, Forum for Energy and Development, and Greenpeace International.

Eyre, N. 1999. Carbon Dioxide Emissions Trends from the United Kingdom. In *Promoting Development While Limiting Greenhouse Gas Emissions: Trends and Baselines.* Edited by J. Goldemberg and W. Reid. New York: United Nations Development Programme.

Fackler, M. 2002. Three Gorges Dam Will Raise Temperatures in Central China, Meteorologist Predicts. Associated Press, April 11.

Fawcett, T. 2001. Retail Therapy: Increasing the Sales of CFLs. In *Proceedings of the 2001 ECEEE Summer Study* 2: 118–129. Paris: European Council for an Energy-Efficient Economy.

Fialka, J. 2001. As the Federal Government Shies Away, States Step Up Efforts to Curb Pollution. *Wall Street Journal*, Sept. 11.

FitzRoy, F. and I. Smith. 1998. Public Transport Demand in Freiburg: Why Did Patronage Double in a Decade? *Transport Policy* 5: 163–174.

Flanigan, T. and P. Rumsey. 1996. Promoting Energy Efficiency in Asia: A Compendium of Asian Success Stories. In *Proceedings of the 1996 ACEEE Summer Study on Energy Efficiency in Buildings* 9: 9.77–9.86. Washington, DC: American Council for an Energy-Efficient Economy.

Flavin, C. and S. Dunn. 1999. Reinventing the Energy System. In *State of the World 1999*, 22–40. Edited by L. Starke. New York: W.W. Norton.

Ford, K. W., G. J. Rochlin, M. H. Ross, and R. H. Socolow. 1975. *Efficient Use of Energy: A Physics Perspective.* College Park, MD: American Institute of Physics.

Friedman, D., J. Mark, P. Monahan, C. Nash, and C. Ditlow. 2001. *Drilling in Detroit: Tapping Automaker Ingenuity to Build Safe and Efficient Automobiles.* Cambridge, MA: Union of Concerned Scientists.

Friedmann, R. 1998. Activities and Lessons Learned Saving Electricity in Mexican Households. In *Proceedings of the 1998 ACEEE Summer Study on Energy Efficiency in Buildings* 5: 119–5:130. Washington, DC: American Council for an Energy-Efficient Economy.

———. 2000. Latin American Experiences with Residential CFL Projects. In *Proceedings of the 2000 ACEEE Summer Study on Energy Efficiency in Buildings* 2: 103–114. Washington, DC: American Council for an Energy-Efficient Economy.

GAO [General Accounting Office]. 2000. *Tax Incentives for Petroleum and Ethanol Fuels.* GAO/RCED-00-301R. Washington, DC: U.S. General Accounting Office. Sept.

Gardiner, D. and L. Jacobson. 2002. Will Voluntary Programs be Sufficient to Reduce U.S. Greenhouse Gas Emissions? An Analysis of the Bush Administration's Global Climate Change Initiative. *Environment 44* (8): 24–33.

Gardner, G. 2001. Accelerating the Shift to Sustainability. In *State of the World 2001,* 189–206. New York: W. W. Norton.

GEF [Global Environmental Facility]. 2001a. *Renewable Energy: GEF Partners with Business for a Better World.* Washington, DC: Global Environmental Facility.

———. 2001b. *Climate Change Program Study Synthesis Report.* Washington, DC: Global Environmental Facility.

GEIPOT. 2001. *Anuario Estatistico dos Transportes. Brasilia: Ministerio dos Transportes.* www.geipot.gov.br.

Geller, H. 1985. Ethanol Fuel from Sugar Cane in Brazil. *Annual Review of Energy* 10: 135–164.

———. 1991. *Efficient Electricity Use: A Development Strategy for Brazil.* Washington, DC: American Council for an Energy-Efficient Economy.

———. 1997. National Appliance Efficiency Standards in the USA: Cost-Effective Federal Regulations. *Energy and Buildings* 26: 101–109.

———. 1999. *Tax Incentives for Innovative Energy-Efficient Technologies.* Washington, DC: American Council for an Energy-Efficient Economy.

———. 2000. *Transforming End-Use Energy Efficiency in Brazil.* Washington, DC: American Council for an Energy-Efficient Economy.

———. 2001. *Strategies for Reducing Oil Imports: Expanding Oil Production vs. Increasing Vehicle Efficiency.* Washington, DC: American Council for an Energy-Efficient Economy.

Geller, H. and R. N. Elliott. 1994. *Industrial Energy Efficiency: Trends, Savings Potential, and Policy Options.* Washington, DC: American Council for an Energy-Efficient Economy.

Geller, H. and D. B. Goldstein. 1998. *Equipment Efficiency Standards: Mitigating Global Climate Change at a Profit.* Washington, DC: American Council for an Energy-Efficient Economy; and San Francisco, CA: Natural Resources Defense Council.

Geller, H. and T. Kubo. 2000. *National and State Energy Use and Carbon Emissions Trends.* Washington, DC: American Council for an Energy-Efficient Economy.

Geller, H. and S. McGaraghan. 1998. Successful Government-Industry Partnership: The U.S. Department of Energy's Role in Advancing Energy-Efficient Technologies. *Energy Policy* 26: 167–177.

Geller, H. and S. Nadel. 1994. Market Transformation Strategies to Promote End-Use Efficiency. *Annual Review of Energy and Environment* 19: 301–346.

Geller, H. and J. Thorne. 1999. *U.S. Department of Energy's Office of Building Technologies: Successful Initiatives of the 1990s.* Washington, DC: American Council for an Energy-Efficient Economy.

Geller, H., J. DeCicco, and S. Nadel. 1993. *Structuring an Energy Tax So That Energy Bills Do Not Increase.* Washington, DC: American Council for an Energy-Efficient Economy.

Geller, H., T. Kubo, and S. Nadel. 2001. *Overall Savings from Federal Appliance and Equipment Efficiency Standards.* Washington, DC: American Council for an Energy-Efficient Economy.

Geller, H., M. Almeida, M. Lima, G. Pimentel, and A. Pinhel. 1999. *Update on Brazil's National Electricity Conservation Program (PROCEL).* Washington, DC: American Council for an Energy-Efficient Economy.

Geller, H., J. P. Harris, M. D. Levine, and A. H. Rosenfeld.1987. The Role of Federal Research and Development in Advancing Energy Efficiency: A $50 Billion Contribution to the U.S. Economy. *Annual Review of Energy* 12: 357–396.

Geller, H., G. M. Jannuzzi, R. Schaeffer, and M. T. Tolmasquim. 1998. The Efficient Use of Electricity in Brazil: Progress and Opportunities. *Energy Policy* 26: 859–872.

Gillespie, M. 2001. Americans Favor Alternative Energy Methods to Solve Shortages. Princeton, N.J.: The Gallup Organization. www.gallup.com/poll/releases/pr011127.asp.

Gipe, P. 2000. Wind Booms Worldwide: Latest BTM Report Paints a Promising Picture. *Renewable Energy World* 3 (4): 132–149.

Goldberg, M. 2000. *Federal Energy Subsidies: Not All Technologies Are Created Equal.* Washington, DC: Renewable Energy Policy Project.

Goldemberg, J. 1998. Leapfrog Energy Technologies. *Energy Policy* 26: 729–742.

———. 1999. *The Issue of Technology Transfer for Development.* São Paulo, Brazil: University of São Paulo.

———. 2000. Rural Energy in Developing Countries. In *World Energy Assessment: Energy and the Challenge of Sustainability.* New York: United Nations Development Programme.

Goldstein, D. B. and A. H. Rosenfeld. 1976. *Projecting an Energy-Efficient California.* LBL-3274/EEB-76-1. Berkeley, CA: Lawrence Berkeley National Laboratory.

Goldstein, L., J. Mortensen, and D. Trickett. 1999. *Grid-Connected Renewable-Electric Policies in the European Union.* NREL/TP.620.26247. Golden, CO: National Renewable Energy Laboratory.

Granda, C. 1997. Case Study: The IFC/GEF Poland Efficient Lighting Project (PELP). In *Proceedings of the 4th European Conference on Energy-Efficient Lighting* 2: 271–277. Frederiksberg, Denmark: DEF Congress Service.

Greene, D. L. 1999. *Why CAFE Worked.* Energy Policy 26: 595–614.

Greene, D. L. and P. N. Leiby. 1993. *The Social Costs to the U.S. of Monopolization of the World Oil Market, 1972–1991.* ORNL-6744. Oak Ridge, TN: Oak Ridge National Laboratory.

Grubler, A. 1998. *Technology and Global Change.* Cambridge: Cambridge University Press.

Grubler, A., N. Nakicenovic, and D. G. Victor. 1999. Dynamics of Energy Technologies and Global Change. *Energy Policy* 27: 247–280.

Gummer, J. and R. Moreland. 2000. *The European Union and Global Climate Change: A Review of Five National Programmes.* Arlington, VA: Pew Center on Global Climate Change.

Hakim, D. 2002. California Is Moving to Guide U.S. Policy on Pollution. *New York Times.* July.

Halverson, M., J. Johnson, D. Weitz, R. Majette, and M. LaLiberte. 2002. Making Residential Energy Codes More Effective: Building Science, Beyond Code Programs, and Effective Implementation Strategies. *Proceedings of the 2002 ACEEE Summer Study on Energy Efficiency in Buildings* 2: 111–22. Washington, DC: American Council for an Energy-Efficient Economy.

Hankins, M. 2001. Commercial Breaks: Building the Market for PV in Africa. *Renewable Energy World* 4 (4): 164–175.

Hawken, P., A. Lovins, and L. H. Lovins. 1999. *Natural Capitalism: Creating the Next Industrial Revolution.* Boston: Little, Brown and Co.

Henriques, M. and R. Schaeffer. 1995. Energy Use in Brazilian Industry: Gains from Energy Efficiency Improvements. In *Proceedings of the ACEEE 1995 Summer Study on Energy Efficiency in Industry* 2: 99–110. Washington, DC: American Council for an Energy-Efficient Economy.

Herzog, H., B. Eliasson, and O. Kaarstad. 2000. Capturing Greenhouse Gases. *Scientific American* 282(2): 72–79.

Hirst, E., E. Blank, and D. Moskovitz. 1994. Alternative Ways to Decouple Electric Utility Revenues from Sales. *Electricity Journal 7* (4): 38–47.

Holdren, J. P. and K. R. Smith. 2000. Energy, the Environment, and Health. In *World Energy Assessment: Energy and the Challenge of Sustainability.* New York: United Nations Development Programme.

Holloway, S. 2001. Storage of Fossil Fuel-Derived Carbon Dioxide Beneath the Surface of the Earth. *Annual Review of Energy and the Environment 26*: 145–66.

Huang, J., J. L. Warner, S. Wiel, A. Rivas, and O. de Buen. 1998. A Commercial Building Energy Standard. In *Proceedings of the 1998 ACEEE Summer Study on Energy Efficiency in Buildings* 5: 153–164. Washington, DC: American Council for an Energy-Efficient Economy.

Hurrell, J. 2002. Personal communication from James Hurrell, National Center for Atmospheric Research, Boulder, CO, September.

Hussain, R. 2001. Micro Credit in Bangladesh. *Sustainable Energy News* 32: 9.

ICLEI [International Council for Local Environmental Initiatives]. 2000a. *U.S. Communities Acting to Protect the Climate.* Berkeley, CA: International Council for Local Environmental Initiatives.

———. 2000b. *Best Practices for Climate Protection: A Local Environmental Guide.* Berkeley, CA: International Council for Local Environmental Initiatives.

IEA [International Energy Agency]. 1997a. *Renewable Energy Policy in IEA Countries.* Paris: International Energy Agency.

———. 1997b. *Enhancing the Market Deployment of Energy Technology: A Survey of Eight Technologies.* Paris: International Energy Agency.

———. 1997c. *Energy Efficiency Initiative. Vol. 2: Country Profiles and Case Studies.* Paris: International Energy Agency.

———. 1997d. *Indicators of Energy Use and Efficiency: Understanding the Link between Energy and Human Activity.* Paris: International Energy Agency.

———. 2000a. *World Energy Outlook 2000.* Paris: International Energy Agency.

———. 2000b. *Energy Labels and Standards.* Paris: International Energy Agency.

———. 2000c. *The Road from Kyoto: Current CO_2 and Transport Policies in the IEA*. Paris: International Energy Agency.
———. 2000d. *Dealing with Climate Change: Policies and Measures in IEA Member Countries.* Paris: International Energy Agency.
———. 2000e. *Experience Curves for Energy Technology Policy.* Paris: International Energy Agency.
———. 2000f. *Energy Policies of IEA Countries: 1999 Review.* Paris: International Energy Agency.
———. 2001a. *End-User Oil Product Prices and Average Crude Oil Import Costs March 2001.* Paris: International Energy Agency.
———. 2001b. *Energy Efficiency Update: United Kingdom.* Paris: International Energy Agency. www.iea.org/pubs/newslett/eneeff/uk.pdf.
———. 2001c. *Energy Efficiency Update: Japan.* Paris: International Energy Agency. www.iea.org/pubs/newslett/eneeff/jp.pdf.
———. 2001d. *About the IEA.* Paris: International Energy Agency. www.iea.org/about
———. 2001e. *Energy Policies of IEA Countries: 2001 Review.* Paris: International Energy Agency.
———. 2001f. *Things That Go Blip in the Night: Standby Power and How to Limit It.* Paris: International Energy Agency.
———. 2001g. *Technology without Borders: Case Studies of Successful Technology Transfer.* Paris: International Energy Agency.
———. 2001h. *Toward a Sustainable Energy Future.* Paris: International Energy Agency.
Interlaboratory Working Group. 1997. *Scenarios of U.S. Carbon Reductions: Potential Impacts of Energy Technologies by 2010 and Beyond.* Oak Ridge, TN: Oak Ridge National Laboratory; and Berkeley, CA: Lawrence Berkeley National Laboratory.
———. 2000. *Scenarios for a Clean Energy Future.* Oak Ridge, TN: Oak Ridge National Laboratory; and Berkeley, CA: Lawrence Berkeley National Laboratory.
IPCC [Intergovernmental Panel on Climate Change]. 2000. *Methodological and Technological Issues in Technology Transfer.* Geneva, Switzerland: Intergovernmental Panel on Climate Change.
———. 2001a. *Climate Change 2001: The Scientific Basis: Summary for Policymakers.* Geneva, Switzerland: Intergovernmental Panel on Climate Change. www.unep.ch/ipcc/pub/spm22-01/pdf.
———. 2001b. *Climate Change 2001: Mitigation. Summary for Policymakers.* Geneva, Switzerland: Intergovernmental Panel on Climate Change. www.unep.ch/ipcc/pub/wg2spmfinal/pdf.

———. 2001c. *Climate Change 2001: Impacts, Adaptation, and Vulnerability: Summary for Policymakers.* Geneva, Switzerland: Intergovernmental Panel on Climate Change. www.unep.ch/ipcc/pub/wg3spm/pdf.

Jacobsson, S. and A. Johnson. 2000. The Diffusion of Renewable Energy Technology: An Analytical Framework and Key Issues for Research. *Energy Policy* 28: 625–640.

Jain, B. C. 2000. Commercialising Biomass Gasifiers: Indian Experience. *Energy for Sustainable Development* 4 (3): 72–83.

Jannuzzi, G. M. 2001. The Prospects for Energy Efficiency, R&D and Climate Change Issues in a Competitive Energy Sector Environment in Brazil. In *Proceedings of the 2001 ECEEE Summer Study* 2: 415–424. Paris: European Council for an Energy-Efficient Economy.

Jensen, M. W. and M. Ross. 2000. The Ultimate Challenge: Developing an Infrastructure for Fuel Cell Vehicles. *Environment* 42 (7): 10–22.

Jessup, P. 2001. *The City of Toronto's Corporate Energy Use and CO_2 Emissions, 1990-98: A Progress Report.* Toronto: Toronto Atmospheric Fund.

Jochem, E. 2000. Energy End-Use Efficiency. In *World Energy Assessment: Energy and the Challenge of Sustainability.* New York: United Nations Development Programme.

Johnson, A. and S. Jacobsson. 2001. *The Emergence of a Growth Industry: A Comparative Analysis of the German, Dutch and Swedish Wind Turbine Industries.* Goteberg, Sweden: Chalmers University of Technology.

Jones, E. and J. Eto. 1997. *Financing End-Use Solar Technologies in a Restructured Utility Industry: Comparing the Cost of Public Policies.* LBNL-40218. Berkeley, CA: Lawrence Berkeley National Laboratory.

Kamalanathan, C. R. 1998. Commercial Wind Power Projects. *IREDA News* 9 (4): 13–16.

Kammen, D. M. 1999. *Building Institutional Capacity for Small-Scale and Decentralized Energy Research, Development, Demonstration, and Deployment (ERD3) in the South.* Berkeley, CA: University of California, Berkeley, Energy and Resources Group.

Karekezi, S. 2002a. Poverty and Energy in Africa—A Brief Review. *Energy Policy* 30: 915–922.

———. 2002b. Renewables in Africa—Meeting the Energy Needs of the Poor. *Energy Policy* 30: 1059–70.

Karekezi, S. and J. Kimani. 2002. Status of Power Sector Reform in Africa: Impact on the Poor. *Energy Policy* 30: 923–46.

Karekezi, S. and T. Ranja. 1997. *Renewable Energy Technologies in Africa.* London: Zed Books.

Kates, R. W. 2000. Population and Consumption: What We Know, What We Need to Know. *Environment* 42 (3): 10–19.

Katsumata, H. 1999. Energy Policy in Japan: New Energy and Renewable Energy. Paper presented at the Energy Expert Network Meeting, Tokyo, Japan, New Energy and Industrial Technology Development Organization.

Kauffman, H. 1999. Johnson & Johnson Strives to Implement Best Practices by 2000. In *Proceedings of the ACEEE 1999 Summer Study on Energy Efficiency in Industry*, 1–12. Washington, DC: American Council for an Energy-Efficient Economy.

Khatib, H. 2000. Energy Security. In *World Energy Assessment: Energy and the Challenge of Sustainability*. New York, NY: United Nations Development Programme.

Kheshgi, H. S., R. C. Prince, and G. Marland. 2000. The Potential for Biomass Fuels in the Context of Global Climate Change: Focus on Transportation Fuels. *Annual Review of Energy and the Environment* 25: 199–244.

Kinney, L. and J. Cavallo. 2000. *Refrigerator Replacement Programs: Putting a Chill on Energy Waste*. ER-00-18. Boulder, CO: Financial Times Energy, Inc.

Krewitt, W., T. Heck, A. Trukenmuller, and R. Friedrich. 1999. Environmental Damage Costs from Fossil Electricity Generation in Germany and Europe. *Energy Policy* 27: 173–183.

Krohn, S. 2002a. *Wind Energy Policy in Denmark: 25 Years of Success—What Now?* Copenhagen: Danish Wind Industry Association. www.windpower.dk/articles/whatnow.htm.

———. 2002b. *Wind Energy Policy in Denmark: Status 2002*. Copenhagen: Danish Wind Industry Association. www.windpower.dk/articles/energypo.htm.

Krugman, P. 2001. Not a Fuels Errand. *New York Times* Sept. 26. www.nytimes.com/2001/09/26/opinion/26KRUG.html.

———. 2002. Ersatz Climate Policy. *New York Times*. Feb. 15. www.nytimes.com/2002/02/15/opinion/15KRUG.html.

Kubo, T., H. Sachs, and S. Nadel. 2001. *Opportunities for New Appliance and Equipment Efficiency Standards: Energy and Economic Savings Beyond Current Standards Programs*. Washington, DC: American Council for an Energy-Efficient Economy.

Kushler, M. and P. Witte. 2001. *A Revised 50-State Status Report on Electric Restructuring and Public Benefits*. Washington, DC: American Council for an Energy-Efficient Economy.

Laitner, S., S. Bernow, and J. DeCicco. 1998. Employment and Other Macroeconomic Benefits of an Innovation-led Climate Strategy for the United States. *Energy Policy* 26: 425–432.

Lal, S. 1998. Renewable Energy Market Development-Initiatives of IREDA. *IREDA News* 9 (4): 39–44.

Lamberts, R. 2001. Personal communication with Roberto Lamberts, Federal University of Santa Catarina, Florianopolis, Brazil.

Lazaroff, C. 2002. BP Reaches Climate Goals Eight Years Early. Environment News Service, March 12. http://ens.lycos.com/ens/mar2002/2002L-03-12-06.html.

Lee, A. D. and R. Conger. 1996. Market Transformation: Does It Work? The Super Efficient Refrigerator Program. In *Proceedings of the 1996 ACEEE Summer Study on Energy Efficiency in Buildings* 3: 69–80. Washington, DC: American Council for an Energy-Efficient Economy.

Lew, D. J., R. H. Williams, X. Shaoxiong, and Z. Shihui. 1998. Large-Scale Baseload Wind Power in China. *Natural Resources Forum* 22 (3): 165–184.

Lew, D. and J. Logan. 2001. The Answer Is Blowin' in the Wind. China Online. www.chinaonline.com. March 12.

Li, J. 1999. Clouds to the East. *Forum for Applied Research and Public Policy* Winter 1999: 55–61.

Lima, J. H. 2002. The Brazilian PRODEEM Programme for Rural Electrification Using Photovoltaics. Presentation at the Rio 02 World Climate and Energy Conference, Rio de Janeiro, January 8–10, 2002.

Loiter, J. M. and V. Norberg-Bohm. 1999. Technology Policy and Renewable Energy: Public Roles in the Development of New Energy Technologies. *Energy Policy* 27: 85–97.

London, S. J. and I. Romieu. 2000. Health Costs due to Outdoor Air Pollution by Traffic. *Lancet* 356: 782.

Lovins, A. B and L. H. Lovins. 1997. *Climate: Making Sense and Making Money.* Old Snowmass, CO: Rocky Mountain Institute.

Lovins, A. B. and B. D. Williams. 1999. *A Strategy for the Hydrogen Transition.* Old Snowmass, CO: Rocky Mountain Institute.

Lutzenhiser, L. and M. H. Grossard. 2000. Lifestyle, Status and Energy Consumption. *Proceedings of the 2000 ACEEE Summer Study on Energy Efficiency in Buildings* 8: 207–222. Washington, DC: American Council for an Energy-Efficient Economy.

Mahlman, J. D. 2001. The Long Time Scales of Human-Caused Climate Warming: Further Challenges for the Global Policy Process. Presentation at the Pew Center on Global Climate Change Workshop on the Timing of Climate Change Policies. Washington, DC, Oct. 11–12, 2001. www.pewclimate.org/events/timing_mahlman.pdf.

Margolick, M. and D. Russell. 2001. *Corporate Greenhouse Gas Reduction Targets.* Arlington, VA: Pew Center on Global Climate Change.

Margolis, R. M. and D. M. Kammen. 1999. Evidence of Under-Investment in Energy R&D in the United States and the Impact of Federal Policy. *Energy Policy* 27: 575–584.

Marsh, P. A. and R. K. Fisher. 1999. It's Not Easy Being Green: Environmental Technologies Enhance Hydropower's Role in Sustainable Development. *Annual Review of Energy and the Environment* 24: 173–188.

Martin, N., E. Worrell, M. Ruth, L. Price, R. N. Elliott, A. M. Shipley, and J. Thorne. 2000. *Emerging Energy-Efficient Industrial Technologies.* Washington, DC: American Council for an Energy-Efficient Economy.

Martinot, E. 1998. Energy Efficiency and Renewable Energy in Russia. *Energy Policy* 26: 905–915.

———. 2001. Renewable Energy Investment by the World Bank. *Energy Policy* 29: 689–699.

Martinot, E. and N. Borg. 1998. Energy-Efficient Lighting Programs. *Energy Policy* 26: 1071–1082.

Martinot, E. and O. McDoom. 2000. *Promoting Energy Efficiency and Renewable Energy: GEF Climate Change Projects and Impacts.* Washington, DC: Global Environmental Facility.

Martinot, E., A. Cabraal, and S. Mathur. 2000. *World Bank/GEF Solar Home Systems Projects: Experiences and Lessons Learned.* Washington, DC: World Bank.

Martinot, E., A. Chaurey, D. Lew, J. Moreira, and N. Wamukonya. 2002. Renewable Energy Markets in Developing Countries. *Annual Review of Energy and the Environment* 27 (forthcoming).

Matrosov, Y. A., M. Chao, and D. B. Goldstein. 2000. Development, Review, and Implementation of Building Energy Codes in Russia: History, Process, and Stakeholder Roles. In *Proceedings of the 2000 ACEEE Summer Study on Energy Efficiency in Buildings* 9: 275–286. Washington, DC: American Council for an Energy-Efficient Economy.

Maycock, P. 2000. The World PV Market 2000: Shifting from Subsidy to "Fully Economic?" *Renewable Energy World* 3 (4): 59–74.

———. 2001. The PV Boom: Where Germany and Japan Lead, Will California Follow? *Renewable Energy World* 4 (4): 144–163.

Mayer, R., E. Blank, and B. Swezey. 1999. *The Grassroots Are Greener: A Community-Based Approach to Marketing Green Power.* Washington, DC: Renewable Energy Policy Project.

McCullough, R. 2001. Price Spike Tsunami: How Market Power Soaked California. *Public Utilities Fortnightly.* Jan. 1.

McDonald, A. and L. Schrattenholzer. 2001. Learning Rates for Energy Technologies. *Energy Policy* 29: 255–261.

McGowan, J. G. and S. R. Connors. 2000. Windpower: A Turn of the Century Review. *Annual Review of Energy and the Environment* 25: 147–198.

McVeigh, D. Burtraw, J. Darmstadter, and K. Palmer. 1999. *Winner, Loser or Innocent Victim: Has Renewable Energy Performed as Expected?* Washington, DC: Renewable Energy Policy Project.

Meyers, E. M. and M. G. Hu 2001. Clean Distributed Generation: Policy Options to Promote Clean Air and Reliability. *Electricity Journal* 14 (1): 89–98.

Meyers, S. 1998. *Improving Energy Efficiency: Strategies for Supporting Sustained Market Evolution in Developing and Transitioning Countries.* LBNL-41460. Berkeley, CA: Lawrence Berkeley National Laboratory.

Mielnik, O. and J. Goldemberg. 2002. Foreign Direct Investment and Decoupling between Energy and Gross Domestic Product in Developing Nations. *Energy Policy* 30: 87–90.

Miller, D. and C. Hope. 2000. Learning to Lend for Off-Grid Solar Power: Policy Lessons from World Bank Loans to India, Indonesia, and Sri Lanka. *Energy Policy* 28: 87–105.

Misana, S. and G. V. Karlsson, eds. 2001. *Generating Opportunities: Case Studies on Energy and Women.* New York: United Nations Development Programme.

Mishra, S. 2000. India Wind Power Rebounding after Late-Nineties Decline. *Clean Energy Finance* 5 (1): 1, 7.

Mitchell, C. 2000. The England and Wales Non–Fossil Fuel Obligation: History and Lessons. *Annual Review of Energy and Environment* 25: 285–312.

MME [Ministry of Mines and Energy]. 1999. *Brazilian Energy Balance 2000.* Brasilia: Ministry of Mines and Energy, Republic of Brazil.

———. 2000. *Brazilian Energy Balance 2001.* Brasilia: Ministry of Mines and Energy, Republic of Brazil. www.mme.gov.br.

MNES [Ministry for Non-Conventional Energy Sources]. 2001. *Annual Report 1999–2000.* New Delhi, India: Ministry for Non-Conventional Energy Sources. www.mnes.nic.in/frame.htm?publications.htm.

Mock, J. E., J. W. Tester, and P. M. Wright. 1997. Geothermal Energy from the Earth: Its Potential Impact as an Environmentally Sustainable Resource. *Annual Review of Energy and the Environment* 22: 305–356.

Moore, C. and J. Ihle. 1999. *Renewable Energy Policy Outside the United States.* Washington, DC: Renewable Energy Policy Project.

Moore, M. 2000. How California Is Advancing Green Power in the New Millennium. *Electricity Journal* 13 (7): 72–77.

Moreira, J. R. 2000. Sugarcane for Energy: Recent Results and Progress in Brazil. *Energy for Sustainable Development* 4 (3): 43–54.

———. 2002. Personal communications with Jose Roberto Moreira, National Biomass Reference Center (CENBIO), São Paulo, July 11, 2002.

Moreira, J. R. and J. Goldemberg. 1999. The Alcohol Program. *Energy Policy* 27: 229–245.

Moriera, J. R., J. Goldemberg, and S. T. Coelho. 2002. *Biomass Availability and Uses in Brazil.* São Paulo: National Biomass Reference Center.

Mulugetta, Y., T. Nhete, and T. Jackson. 2000. Photovoltaics in Zimbabwe: Lessons from the GEF Solar Project. *Energy Policy* 28: 1069–1080.

Murtishaw, S. and L. Schipper. 2001. Disaggregated Analysis of U.S. Energy Consumption in the 1990s: Evidence of the Effects of the Internet and Rapid Economic Growth. *Energy Policy* 29: 1335–56.

Nadel, S. 2002. Appliance and Equipment Efficiency Standards. *Annual Review of Energy and the Environment* 27 (forthcoming).

Nadel, S. and H. Geller. 1996. Utility DSM: What Have We Learned? Where are We Going? *Energy Policy* 24: 289–302.

———. 2001. *Smart Energy Policies: Saving Money and Reducing Pollutant Emissions Through Greater Energy Efficiency.* Washington, DC: American Council for an Energy-Efficient Economy.

Nadel, S. and M. Kushler. 2000. Public Benefits Funds: A Key Strategy for Advancing Energy Efficiency. *Electricity Journal* 13 (8): 74–84.

Nadel, S. and L. Latham. 1998. *The Role of Market Transformation Strategies in Achieving a More Sustainable Energy Future.* Washington, DC: American Council for an Energy-Efficient Economy.

Nadel, S., J. Lin, Y. Cong, A. Hinge, and L. Wenbin. 1999. *The China Green Lights Program: A Status Report.* Washington, DC: American Council for an Energy-Efficient Economy.

Nadel, S., L. Ranier, M. Shepard, M. Suozzo, and J. Thorne. 1998. *Emerging Energy-Saving Technologies and Practices for the Buildings Sector.* Washington, DC: American Council for an Energy-Efficient Economy.

Nadel, S., W. Wanxing, P. Liu, and A. McKane. 2001. The China Motor Systems Energy Conservation Program: A Major National Initiative to Reduce Motor System Energy Use. In *Proceedings of the 2001 ACEEE Summer Study on Energy Efficiency in Industry* 2: 399–413. Washington, DC: American Council for an Energy-Efficient Economy.

Nakata, T. and A. Lamont. 2001. Analysis of the Impacts of Carbon Taxes on Energy Systems in Japan. *Energy Policy* 29: 159–166.

Nakicenovic, N. 2000. Energy Scenarios. In *World Energy Assessment: Energy and the Challenge of Sustainability.* New York, NY: United Nations Development Programme.

Nakicenovic, N., A. Grubler, and A. McDonald. 1998. *Global Energy Perspectives.* Cambridge, U.K: Cambridge University Press.

NARUC [National Association of Regulatory Commissioners]. 1988. *Least Cost Utility Planning Handbook for Public Utilities Commissioners.* Washington, DC: National Association of Regulatory Commissioners.

NAS [National Academy of Sciences]. 1999. *Our Common Journey: A Transition Toward Sustainability.* Washington, DC: National Academy Press.

———. 2001a. *Energy Research at DOE: Was It Worth It?* Washington, DC: National Academy Press.

———. 2001b. *Effectiveness and Impact of Corporate Average Fuel Economy (CAFE) Standards.* Washington, DC: National Academy Press.

NCLC [National Consumer Law Center]. 1995. *Energy and the Poor: The Crisis Continues.* Boston, MA: National Consumer Law Center.

NEEA [Northwest Energy Efficiency Alliance]. 2002. Sales of Efficient Lightbulbs a Bright Spot in 2001. Press Release, April 23, 2002. Portland, OR: Northwest Energy Efficiency Alliance.

Neij, L. 2001. Methods for Evaluating Market Transformation Programmes: Experience in Sweden. *Energy Policy* 29: 67–79.

Neme, C., J. Proctor, and S. Nadel. 1999. *Energy Savings Potential from Addressing Residential Air Conditioner and Heat Pump Installation Problems.* Washington, DC: American Council for an Energy-Efficient Economy.

NEPDG [National Energy Policy Development Group]. 2001. *National Energy Policy.* Washington, DC: U.S. Government Printing Office.

NEPO [National Energy Policy Office]. 2002. *Strategic Plan for Energy Conservation in the Period 2002–2011.* Bangkok, Thailand: National Energy Policy Office.

New Energy Plaza 2001. Result of the Subsidy Program for Residential PV Systems. *New Energy Plaza* 16 (3): 13. Tokyo: New Energy Foundation.

Newman, J. 1998. Evaluation of Energy-Related Voluntary Agreements. In *Industrial Energy Efficiency Policies: Understanding Success and Failure.* Edited by N. Martin, E. Worrell, A. Sandoval, J. W. Bode, and D. Phylipsen. LBNL-42368. Berkeley, CA: Lawrence Berkeley National Laboratory.

NHA [National Hydrogen Association]. 2001. *Hydrogen: The Common Thread: Proceedings of the 12th Annual U.S. Hydrogen Meeting.* Washington, DC: National Hydrogen Association.

Nilsson, L., S. Thomas, C. Lopes, and L. Pagliano. 2001. Energy Efficiency Policy in Restructuring European Electricity Markets. In *Proceedings of the 2001 ECEEE Summer Study* 2: 298–309. Paris: European Council for an Energy-Efficient Economy.

NJDEP [New Jersey Department of Environmental Protection]. 2000. *Global Cli-*

mate Change and Greenhouse Gases. Trenton, NJ: Department of Environmental Protection. www.state.nj.us/dep/dsr/gcc/gcc.htm.

Nogee, A., S. Clemmer, B. Paulos, and B. Haddad. 1999. *Powerful Solutions: Seven Ways to Switch America to Renewable Electricity.* Cambridge, MA: Union of Concerned Scientists.

NPPC [Northwest Power Planning Council] 1998. *Revised Fourth Northwest Conservation and Electric Power Plan.* Portland, OR: Northwest Power Planning Council.

———. 2001. *An Efficiency Power Plant in Three Years: An Interim Goal for the Northwest.* Portland, OR: Northwest Power Planning Council.

NRDC [Natural Resources Defense Council]. 2001. *Slower, Costlier and Dirtier: A Critique of the Bush Energy Plan.* Washington, DC: Natural Resources Defense Council.

NRDC [Natural Resources Defense Council] and SVMG [Silicon Valley Manufacturing Group]. 2001. *Energy Efficiency Leadership in a Crisis: How California Is Winning.* San Francisco: Natural Resources Defense Council and the Silicon Valley Manufacturing Group.

Nuijen, W.C. 1998. Long-Term Agreements on Energy Efficiency in Industry. In *Industrial Energy Efficiency Policies: Understanding Success and Failure,* 79–90. Edited by N. Martin, E. Worrell, A. Sandoval, J. W. Bode, and D. Phylipsen. LBNL-42368. Berkeley, CA: Lawrence Berkeley National Laboratory.

NYS. 2001. Governor Announces Creation of Greenhouse Gas Task Force. Press Release. June 10. Albany, NY: Office of the Governor. www.state.ny.us/governor/press/year01/june10_01/htm.

NYSEPB [New York State Energy Planning Board]. 2002. *2002 State Energy Plan.* Albany, NY: New York State Energy Planning Board.

Odgaard, O. 2000. *Renewable Energy in Denmark.* Copenhagen: Danish Energy Agency.

Ogden, J. M., R. H. Williams, and E. D. Larson. 2001. *Toward a Hydrogen-Based Transportation System.* Princeton, NJ: Princeton University, Center for Energy and Environmental Studies.

Oliveira, A. et. al. 1998. *Energia e Desenvolvimento Sustentavel.* Rio de Janeiro: Universidade Federal do Rio de Janeiro, Instituto de Economia.

Oliver, M. and T. Jackson. 1999. The Market for Solar Photovoltaics. *Energy Policy* 27: 371–385.

O'Neill, B.C. and M. Oppenheimer. 2002. Climate Change: Dangerous Climate Impacts and the Kyoto Protocol. *Science* 296(5575): 1971–72.

Osborn, J., C. Goldman, N. Hopper, and T. Singer. 2002. *Assessing U.S. ESCO In-*

dustry Performance and Market Trends: Results from the NAESCO Database Project. LBNL-50304. Berkeley, CA: Lawrence Berkeley National Laboratory.

Overend, R. 2002. Personal communication with Ralph Overend, National Renewable Energy Laboratory, Golden, CO, May.

Pachauri, R. K. and S. Sharma. 1999. India's Achievements in Energy Efficiency and Reducing CO_2 Emissions. In *Promoting Development While Limiting Greenhouse Gas Emissions: Trends and Baselines.* Edited by J. Goldemberg and W. Reid. New York, NY: United Nations Development Programme.

Padmanabhan, S. 1999. (Asia Alternative Energy Unit, The World Bank). Personal communication. March 5.

Pavan, M. 2002. What's Up in Italy? Market Liberalization, Tariff Regulation and Incentives to Promote Energy Efficiency in End-Use Sectors. *Proceedings of the 2002 ACEEE Summer Study on Energy Efficiency in Buildings* 5: 259–70. Washington, DC: American Council for an Energy-Efficient Economy.

Payne, A., R. Duke, and R. H. Williams. 2001. Accelerating Residential PV Expansion: Supply Analysis for Competitive Electricity Markets. *Energy Policy* 29: 787–800.

[PCAST] President's Committee of Advisors on Science and Technology. 1997. *Federal Energy Research and Development for the Challenges of the Twenty-First Century.* Washington, DC: Executive Office of the President, President's Committee of Advisors on Science and Technology, Panel on Energy Research and Development,

———. 1999. *Report to the President on the Federal Role in International Cooperation on Energy Innovation.* Washington, DC: President's Committee of Advisors on Science and Technology.

Petkova, E. and K. A. Baumert. 2000. *Managing Joint Implementation Work: Lessons from Central and Eastern Europe.* Washington, DC: World Resources Institute.

Phillips, K. 2002. The Company Presidency. *Los Angeles Times* Feb. 20.

Phylipsen, D., K. Blok, E. Worrell, and J. de Beer. 2002. Benchmarking the Energy Efficiency of Dutch Industry: An Assessment of the Expected Effect on Energy Consumption and CO_2 Emissions. *Energy Policy* 30: 663–79.

Phylipsen, D., L. Price, E. Worrell, and K. Blok. 1999. Industrial Energy Efficiency in Light of Climate Change Negotiations: Comparing Major Developing Countries and the U.S. In *Proceedings of the 1999 ACEEE Summer Study on Energy Efficiency in Industry* 193–207. Washington, DC: American Council for an Energy-Efficient Economy.

PNAD 1999. *Pesquisa Nacional por Amostra de Domicilios.* Rio de Janeiro: Instituto Brasileiro de Geografia e Estatistica.

Pollard, V. 2001. Wind in Europe: Developments in the Policy Frameworks for Wind Energy. *Renewable Energy World* 4 (4): 88–101.

Price, L. and E. Worrell. 2000. *International Industrial Sector Energy Efficiency Policies.* LBNL-46274. Berkeley, CA: Lawrence Berkeley National Laboratory.

Quinlan, P., H. Geller, and S. Nadel. 2001. *Tax Incentives for Innovative Energy-Efficient Technologies (Updated).* Washington, DC: American Council for an Energy-Efficient Economy.

Rabinovitch, J. and J. Leitman. 1996. Urban Planning in Curitiba. *Scientific American* March: 46–53.

Rajsekhar, B., F. van Hulle, and J. C. Jansen. 1999. Indian Wind Energy Programme: Performance and Future Directions. *Energy Policy* 27: 669–678.

Ramakrishna, K. and O. R. Young. 1997. International Organizations in a Warming World: Building a Global Climate Regime. In *South–South North Partnership on Climate Change and Greenhouse Gas Emissions,* 13–36. Edited by S. K. Ribeiro and L. P. Rosa. Rio de Janeiro, Brazil: COPPE, Federal University of Rio de Janeiro.

Raskin, P., T. Banuri, G. Gallopin, P. Gutman, A. Hammond, R. Kates, and R. Swart. 2002. *Great Transition: The Promise and Lure of the Times Ahead.* Boston: Stockholm Environment Institute–Boston.

Ravindranath, N. H. and D. O. Hall. 1995. *Biomass, Energy and Development: A Developing Country Perspective from India.* Oxford and New York: Oxford University Press.

Ray, P. H. and S. R. Anderson. 2000. *The Cultural Creatives.* New York: Harmony Books.

Reddy, A. K. N. 1991. Barriers to Improvements in Energy Efficiency. *Energy Policy* 19: 953–961.

———. 2000. Energy and Social Issues. In *World Energy Assessment: Energy and the Challenge of Sustainability.* New York, NY: United Nations Development Programme.

———. 2001. Indian Power Sector Reform for Sustainable Development: The Public Benefits Imperative. *Energy for Sustainable Development* 4 (2): 74–81.

Reddy, A. K. N., Y. P. Anand, and A. D'Sa. 2000. Energy for a Sustainable Road/Rail Transport System in India. *Energy for Sustainable Development* 4 (1): 29–44.

Reddy, A. K. N., R. H. Williams, and T. B. Johansson. 1997. *Energy after Rio: Prospects and Challenges.* New York, NY: United Nations Development Programme.

Reid, W. V. and J. Goldemberg. 1998. Developing Countries Are Combating Climate Change. *Energy Policy* 26: 233–237.

Renner, M. 2000. *Working for the Environment: A Growing Source of Jobs.* Washington, DC: Worldwatch Institute. Sept.

Rever, B. 2001. Grid-Tied Markets for Photovoltaics: A New Source Emerges. *Renewable Energy World* 4 (4): 176–189.

Rietbergen, M., J. Farla, and K. Blok. 1998. Quantitative Evaluation of Voluntary Agreements on Energy Efficiency. In *Industrial Energy Efficiency Policies: Understanding Success and Failure,* 63–78. Edited by N. Martin, E. Worrell, A. Sandoval, J. W. Bode, and D. Phylipsen. LBNL-42368. Berkeley, CA: Lawrence Berkeley National Laboratory.

Rogner, H.-H. 2000. Energy Resources. In *World Energy Assessment: Energy and the Challenge of Sustainability.* New York: United Nations Development Programme.

Rogner, H.-H. and A. Popescu. 2000. An Introduction to Energy. In *World Energy Assessment: Energy and the Challenge of Sustainability.* New York: United Nations Development Programme.

Roizenblatt, I. 2002. Personal communication with Isac Roizenblatt, Philips Lighting of Brazil, São Paulo, Brazil. Feb.

Romm, J., A. Rosenfeld, and S. Herrmann. 1999. *The Internet Economy and Global Warming.* Annandale, VA: Center for Energy and Climate Solutions.

Roodman, D. M. 1998. *The Natural Wealth of Nations: Harnessing the Market for the Environment.* New York: W.W. Norton.

Rosillo-Calle, F. and L. Cortez 1998. Towards ProAlcool II: A Review of the Brazilian Bioethanol Programme. *Biomass and Bioenergy Policy* 14: 115–124.

Ross, M. H. and R. H. Williams. 1977. The Potential for Fuel Conservation. *Technology Review* 79: 49–58.

———. 1981. *Our Energy: Regaining Control.* New York: McGraw-Hill.

Sadler, S. 1999. *Oregon Carbon Dioxide Emission Standards for New Energy Facilities.* Salem, OR: Department of Consumer and Business Services, Office of Energy.

Sathaye, J. A. and N. H. Ravindranath. 1998. Climate Change Mitigation in the Energy and Forestry Sectors of Developing Countries. *Annual Review of Energy and the Environment* 23: 387–438.

Sawin, J. L. 2002. *Losing the Clean Energy Race: How the U.S. Can Retake the Lead and Solve Global Warming.* Washington, DC: Greenpeace USA.

Schaeffer, R. and A. S. Szklo. 2001. Future Electric Power Technology Choices in Brazil: A Possible Conflict between Local Pollution and Global Climate Change. *Energy Policy* 29: 355–370.

Schipper, L. 1991. *Life-Styles and Energy: A New Perspective.* Berkeley, CA: Lawrence Berkeley National Laboratory.

Schipper, L., R. B. Howarth, and H. Geller. 1990. United States Energy Use from

1973 to 1987: The Impacts of Improved Efficiency. *Annual Review of Energy* 15: 455–504.

Schipper, L., F. Unander, S. Murtishaw, and M. Ting. 2001. Indicators of Energy Use and Carbon Emissions: Explaining the Energy Economy Link. *Annual Review of Energy and the Environment* 26: 49–82.

Schneider, S. H. and C. Azar. 2001. Are Uncertainties in Climate Change and Energy Systems a Justification for Stronger Near-Term Mitigation Policies? Presentation at the Pew Center on Global Climate Change Workshop on the Timing of Climate Change Policies. Washington, DC, Oct. 11–12, 2001. www.pewclimate.org/events/timing_azar_schneider.pdf.

Scholand, M. 2002. Compact Fluorescents Set Record. In *Vital Signs 2002*, 46–47. Edited by L. Starke. New York: W. W. Norton.

Scullion, M. 2001. *Digest of United Kingdom Energy Statistics 2001.* London: Stationary Office. www.dti.gov.uk/epa/digest01.

Seattle 2001. Climate Change. www.cityofseattle.net/light/climatechange.

Shailaja, R. 2000. Women, Energy and Sustainable Development. *Energy for Sustainable Development* 4 (1): 45–64.

Sheehan, M. O. 2001. Making Better Transportation Choices. In *State of the World 2001*, 103–122. Edited by L. Starke. New York: W. W. Norton.

Shepard, M. 2001. *Green Money: Compensating Efficiency and Renewable Energy for their Environmental Benefits.* Boulder, CO: E Source.

Shoda, T. 1999. Outlook for Introduction of Renewable Energy Sources in Japan. *Energy Policy* 27 (1): 57–68.

Shoengold, D. 2001. Personal communication based on data reported by utilities to the Federal Energy Regulatory Commission in FERC Forms 759, 767, and 861. Middleton, WI: MSB Energy Associates.

Shorey, E. and T. Eckman. 2000. *Appliances and Global Climate Change.* Arlington, VA: Pew Center on Global Climate Change.

Short, W. 2002. Renewable Energy Technologies: Progress, Markets, and Industries. Presentation at the 2nd Renewable Energy Analysis Forum, National Renewable Energy Laboratory, Golden, CO, May 29, 2002.

Shukla, P.R., D. Ghosh, W. Chandler, and J. Logan. 1999. *Developing Countries and Global Climate Change: Electric Power Options in India.* Arlington, VA: Pew Center on Global Climate Change.

Simm, I., A. Haq, and V. Widge. 2000. Solar Home Systems in Kenya. *Renewable Energy World* 3 (6): 46–53.

Singh, V. 2001. *The Work That Goes Into Renewable Energy.* Washington, DC: Renewable Energy Policy Project. Nov.

Sinton, J. E. and D. G. Fridley 2000. What Goes Up: Recent Trends in China's Energy Consumption. *Energy Policy* 28: 671–687.

Sinton, J. E., M. D. Levine, and W. Qingyi. 1998. Energy Efficiency in China: Accomplishments and Challenges. *Energy Policy* 26: 813–830.

Smil, V. 2000. Energy in the Twentieth Century: Resources, Conversions, Costs, Uses, and Consequences. *Annual Review of Energy and the Environment* 25: 21–51.

Smith, K. R., G. Shuhua, H. Kun, and Q. Daxiong. 1993. One Hundred Million Improved Cookstoves in China: How Was It Done? *World Development* 21: 941–961.

Soares, J. B., A. S. Szklo, and M. T. Tolmasquim. 2001. Incentive Policies for Natural Gas-fired Cogeneration in Brazil's Industrial Sector—Case Studies: Chemical Plant and Pulp Mill. *Energy Policy* 29: 205–215.

Socolow, R. H. 1977. The Coming Age of Conservation. *Annual Review of Energy* 2: 239–289.

Spalding-Fecher, R., A. Williams, and C. van Horen. 2000. Energy and Environment in South Africa: Charting a Course to Sustainability. *Energy for Sustainable Development* 4 (4): 8–17.

Sperling, D. and D. Salon. 2002. *Transportation in Developing Countries: An Overview of Greenhouse Gas Reduction Strategies.* Arlington, VA: Pew Center on Global Climate Change.

Stoft, S., J. Eto, and S. Kito. 1995. *DSM Shareholder Incentives: Current Design and Economic Theory.* LBL-36580. Berkeley, CA: Lawrence Berkeley National Laboratory.

Strachan, N. and H. Dowlatabadi. 2002. Distributed Generation and Distributed Utilities. *Energy Policy* 30: 649–61.

Streets, D. G., K. Jiang, X. Hu, J. E. Sinton, X.-Q. Zhang, D. Xu, M. Z. Jacobson, and J. E. Hansen. 2001. Recent Reductions in China's Greenhouse Gas Emissions. *Science* 294: 1835–1836.

Strickland, C. and R. Sturm. 1998. Energy Efficiency in World Bank Power Sector Policy and Lending: New Opportunities. *Energy Policy* 26: 873–884.

Suarez, C. E. 1999. Argentina's Ongoing Efforts to Lower Greenhouse Gas Emissions. In *Promoting Development While Limiting Greenhouse Gas Emissions: Trends and Baselines.* Edited by J. Goldemberg and W. Reid. New York: United Nations Development Programme.

Suozzo, M. and J. Thorne. 1999. *Market Transformation Initiatives: Making Progress.* Washington, DC: American Council for an Energy-Efficient Economy.

Swisher, J. N., G. M. Jannuzzi, and R. Y. Redlinger. 1997. *Tools and Methods for Integrated Resource Planning: Improving Energy Efficiency and Protecting the Environment.* Roskilde, Denmark: UNEP Collaborating Centre on Energy and Environment.

Tabosa, R. 1999. Personal communication with Ronaldo Tabosa, PROCEL, Eletrobrás, Rio de Janeiro, Brazil.

Tellus Institute 1999. *Putting the Brakes on Sprawl: Innovative Transportation Solutions from the U.S. and Europe.* Boston: Tellus Institute.

TERI [Tata Energy Research Institute]. 2002. *Renewable Energy Power: An Indian Perspective.* New Delhi: Tata Energy Research Institute. www.teriin.org/opet/articles/art10.htm

Thigpen, S., A. Fanara, A. ten Cate, P. Bertoldi, and T. Takigawa. 1998. Market Transformation Through International Cooperation: The ENERGY STAR® Office Equipment Example. In *Proceedings of the 1998 ACEEE Summer Study on Energy Efficiency in Buildings.* Washington, DC: American Council for an Energy-Efficient Economy.

Thorne, S. and E. L. LaRovere. 1999. Criteria and Indicators for Appraising Clean Development Mechanism (CDM) Projects. Paris: Helio International.

Timilsina, G., T. Lefevre, and S. K. N. Uddin. 2001. New and Renewable Energy Technologies in Asia. *Renewable Energy World* 4 (4): 52–67.

Tolmasquim, M. T. 2001. As Origens da Crise Energetica Brasileira. *Revista Ambiente e Sociedade* 3 (6/7): 179–184.

Tolmasquim, M. T., L. P. Rosa, A. S. Szklo, M. Schuller, and M. A. Delgado. 1998. *Tendencias da Eficiencia Eletrica no Brasil.* Rio de Janeiro: ENERGE/Eletrobrás.

Tolmasquim, M. T. and A. S. Szklo. 2000. *A Matriz Energetica Brasileira na Virada do Milenio.* Rio de Janeiro: COPPE/UFRJ and ENERGE.

Turkenburg, W. C. 2000. Renewable Energy Technologies. In *World Energy Assessment: Energy and the Challenge of Sustainability.* New York: United Nations Development Programme.

UCCEE [UNEP Collaborating Centre on Energy and Environment]. 2001. UCCEE–UNEP Collaborating Centre on Energy and Environment. Roskilde, Denmark: UNEP Collaborating Centre on Energy and Environment. www.uccee.org/about.htm.

UCS [Union of Concerned Scientists]. 2001. *Renewable Portfolio Standards at Work in the States.* Cambridge, MA: Union of Concerned Scientists. www.ucsusa.org/energy/fs_state_rps.html.

———. 2002. *Energy Security: Solutions to Protect America's Power Supply and Reduce Oil Dependence.* Cambridge, MA: Union of Concerned Scientists.

UNDP [United Nations Development Programme]. 2000. *World Energy Assessment: Energy and the Challenge of Sustainability.* New York: United Nations Development Programme.

———. 2001. UNDP Energy for Sustainable Development. New York: United

Nations Development Programme. www.undp.org/seed/eap/About/About.html.

UNFPA [United Nations Population Fund]. 2001. *The State of World Population 2001.* New York: United Nations Population Fund.

Urbanchuk, J. M. 2001. *An Economic Analysis of Legislation for a Renewable Fuels Requirement for Highway Motor Fuels.* Washington, DC: Renewable Fuels Association.

Van Luyt, P. 2001. LTA's and the Recent Covenant Benchmarking Energy Efficiency Agreements in the Netherlands. Presentation at the IEA Workshop on Government-Industry Cooperation to Improve Energy Efficiency and the Environment through Voluntary Action, Washington, DC, Feb. 22. www.iea.org/workshop/gov/govpvlf.pdf.

Vehmas, J., J. Kaivo-oja, J. Luukkanen, and P. Malaska. 1999. Environmental Taxes on Fuels and Electricity–Some Experiences from the Nordic Countries. *Energy Policy* 27: 343–355.

Verani, A., C. Nielsen, and P. Covell. 1999. PV Powers Rural Communities. *Solar Today* May/June: 30–33.

Villaverde, V. 2001. Personal communication with Victor Villaverde, PROCEL, Eletrobrás, Rio de Janeiro. Nov.

Vine, E. 2000. Promoting Emerging Energy-Efficiency Technologies and Practices by Utilities in a Restructured Energy Industry: A Report from California. In *Proceedings of the 2000 ACEEE Summer Study on Energy Efficiency in Buildings* 9: 383–394. Washington, DC: American Council for an Energy-Efficient Economy.

Vine, E. and D. Crawley. 1991. *State of the Art of Energy Efficiency: Future Directions.* Washington, DC: American Council for an Energy-Efficient Economy.

Vivier, J. and M. Mezghani. 2001. The Millennium Cities Database: A Tool for Sustainable Mobility. In *Proceedings of the 2001 ECEEE Summer Study* 1: 474–479. Paris: European Council for an Energy-Efficient Economy.

Volpi, G. 2000. Taking the Road to Renewables? Strengths and Weaknesses of the Draft European Renewables Directive. *Renewable Energy World* 3 (6): 90–97.

Vongsoasup, S., P. Sinsukprasert, P. du Pont, and T. Hernoe. 2002. Piloting the Way to a More Effective Energy Strategy: Thailand's Simplified Subsidy and Finance Initiatives. In *Proceedings of the ACEEE 2002 Summer Study on Energy Efficiency in Buildings* 4: 351–364. Washington, DC: American Council for an Energy-Efficient Economy.

Wachsmann, U. and M. T. Tolmasquim. 2002. Windpower in Brazil: A Transition Using the German Experience. Presentation at the Rio-02 World Climate and Energy Conference. Rio de Janeiro, January 8–10, 2002.

Wagner, A. 2000. Set for the 21st Century: Germany's New Renewable Energy Law. *Renewable Energy World* 3 (2): 73–83.

Waide, P. 2001. Findings of the Cold II SAVE Study to Revise Cold Appliance Energy Labeling and Standards in the EU. In *Proceedings of the 2001 ECEEE Summer Study* 2: 376–389. Paris: European Council for an Energy-Efficient Economy.

Watson, R. T., J. A. Dixon, S. P. Hamburg, A. C. Janetos, and R. H. Moss. 1998. *Protecting Our Planet Securing Our Future: Linkages among Global Environmental Issues and Human Needs.* Nairobi: United Nations Environment Programme.

Weiss, I. and P. Sprau. 2002. 100,000 roofs and 99 Pfenning: Germany's PV Financing Schemes and the Market. *Renewable Energy World* 5 (1).

WHO 1997. *Air Quality Guidelines.* Washington, DC: World Health Organization.

Wiel, S. and J. E. McMahon. 2001. *Energy-Efficiency Labels and Standards: A Guidebook for Appliances, Equipment, and Lighting.* Washington, DC: Collaborative Labeling and Appliance Standards Program.

Wiel, S., J. Busch, C. Sanchez, J. Deringer, E. Fernandez, and M. Companano. 1998. Implementing Energy Standards for Motors and Buildings in the Philippines. In *Proceedings of the 1998 ACEEE Summer Study on Energy Efficiency in Buildings* 5: 339–5:350. Washington, DC: American Council for an Energy-Efficient Economy.

Wilkinson, C. F. 1992. *Crossing the Next Meridian: Land, Water, and the Future of the West.* Washington, DC: Island Press.

Williams, R. H 2000. Advanced Energy Supply Technologies. In *World Energy Assessment: Energy and the Challenge of Sustainability.* New York: United Nations Development Programme.

———. 2002. *Facilitating Widespread Deployment of Wind and Photovoltaic Technologies.* Princeton, NJ: Princeton University, Center for Energy and Environmental Studies.

Winrock International 1999. *Trade Guide on Renewable Energy in Brazil.* Washington, DC: U.S. Agency for International Development.

Wooley, D. R. 2000. A *Guide to the Clean Air Act for the Renewable Energy Community.* Washington, DC: Renewable Energy Policy Project.

World Bank, The. 1997. *Can the Environment Wait? Priorities for East Asia.* Washington, DC: World Bank.

———. 1998. *Fuel for Thought: A New Environmental Strategy for the Energy Sector.* October 22 draft. Washington, DC: World Bank, Environment and Energy, Mining and Telecommunications Departments.

———. 2000. The World Bank Asia Alternative Energy Program (ASTAE) Status

Report #8. Washington, DC: World Bank. www.worldbank.org/astae/astaestatusreport8.pdf.

Worrell, E., N. Martin, and L. Price. 1999. *Energy Efficiency and Carbon Dioxide Emissions Reduction Opportunities in the U.S. Iron and Steel Industry.* LBNL-41724. Berkeley, CA: Lawrence Berkeley National Laboratory.

WRI [World Resources Institute]. 1998. *World Resources 1998–99.* New York: Oxford University Press.

Wright, L. L. and L. A. Kszos. 1999. Bioenergy Status and Expansion in the United States. *Proceedings of the Third IEA Bioenergy Meeting.* Paris: International Energy Agency.

Wyman, C. E. 1999. Biomass Ethanol: Technical Progress, Opportunities, and Commercial Challenges. *Annual Review of Energy and the Environment* 24: 189–226.

Xcel Energy 2001. Denver: Colorado's Second Commercial Wind Farm Debuts. Press Release. Xcel Energy. www.xcelenergy.com/NewsRelease/news Release101601.asp.

Yergin, D. 1991. *The Prize: The Epic Quest for Oil, Money and Power.* New York: Simon and Schuster.

Zhang, C., M. M. May, and T. C. Heller. 2001. Impact of Global Warming of Development and Structural Changes in the Electricity Sector of Guangdong Province, China. *Energy Policy* 29: 179–204.

Zhang, Z. X. 1999. Is China Taking Actions to Limit Its Greenhouse Gas Emissions? Past Evidence and Future Prospects. In *Promoting Development While Limiting Greenhouse Gas Emissions: Trends and Baselines*, 45–58. Edited by J. Goldemberg and W. Reid. New York: United Nations Development Programme.

Index

Acid rain, 6
Africa, sub-Saharan:
  electricity use, 15
  hydropower, 21
  inequalities, consumption, 14, 15
  information and training, 70
  international policies/institutions, 194, 201
  quality problems as barrier to renewable energy use, 41
  solar photovoltaic power, 41
  see also individual countries
Agricultural residues burned for heating/cooking, 7
Air pollution:
  Brazil, 171
  ethanol, 120–21
  market obligations, 81
  trends in, local/regional, 6–8
  see also Carbon dioxide
Alcoa, 127
Annual Energy Outlook 2001, 136–37, 149
Appliances, 51, 97–100, 143–44, 172–73
Arctic National Wildlife Refuge (ANWR), 12, 140, 158
Argentina, market reforms in, 75, 77
Asia, consumption inequalities in East, 14
Asia, Southeast:
  air pollution, 6
  efficiency, energy, 24
  inequalities, consumption, 14–15
  international policies/institutions, 203
  see also China; India
Audits, energy, 35
Australia, market obligations in, 79
Austria:
  air pollution, 7
  plans, energy, 84
Automobiles:
  as barrier to energy efficiency, 40
  Brazil, 178
  future, towards a sustainable energy, 233–34
  market obligations, 80–81
  market transformation, 100–102
  plans, energy, 85
  regulations, 69
  taxes, 63–64
  United States, 131–32

Bagasse cogeneration systems, 179–81
Ballasts/lamps, 51
  see also Compact fluorescent lamps
Bangladesh:
  financing, 53
  population figures/issues, 237
Banks, multilateral development, 202–3

Barriers to greater energy efficiency:
incentives, financial, 36
information and training, 35–36
money/financing, 37–38
overview, 33, 35
political obstacles, 40
pricing and tax barriers, 38–39
purchasing procedures, 36–37
quality problems, 34–35
regulatory and utility, 39–40
subsidies, 38–39
summary/review, 44–45
supply infrastructure, 34
see also Market transformation approach
Barriers to renewable energy use:
financing, 42
information and training, 41–42
overview, 33, 35
political obstacles, 43–44
pricing and tax barriers, 42–43
quality problems, 41
regulatory and utility barriers, 43
subsidies, 42–43, 59–60
summary/review, 44–45
supply infrastructure, 40–41
Base/Clean Energy Scenarios for Brazil, 183–87
Base *vs.* Clean Energy Scenario, U.S., 148–60
Bilateral assistance programs, 200–202, 210, 220
bin Laden, Osama, 12
Bioethanol, 148
Biomass energy:
Brazil, 166
Clean Energy Scenario, Global, 231–32
future, towards a sustainable energy, 221
global energy supply, 3
historical look at energy use, 2
incentives, financial, 55–56
information/training as barrier to renewable energy use, 42
international policies/institutions, 194
market transformation, 115
quality problems as barrier to renewable energy use, 41
research and development, 50
status of the technology, 19–21
Sweden, 20
taxes, 61–62
Boehlart, Sherwood, 147
Bonneville Power Administration, 87
Brazil:
Base/Clean Energy Scenarios, 183–87
biomass energy, 221
capacity building, 82
ethanol fuel, 117–22, 195
evolution of different energy sources, 166–69
financing, 54
incentives, financial, 55
information and training, 71
market reforms, 76
market transformation, 102–6, 117–22
objectives and interests, 169–72
plans, energy, 85
policy initiatives/options, 172–83
population figures/issues, 166
privatization/restructuring, 28
summary/review, 187–89
taxes, 62
British Petroleum (BP), 127
Builders as barriers to energy efficiency, 36
Building energy standards, 67–68, 112, 144, 174–75
Bush, George H. W., 135–37, 204
Bush, George W., 143, 162
Business-as-usual base scenario, 135–37
Businesses as barrier to energy efficiency, 37, 40

California:
electricity, 28
future, towards a sustainable energy, 223–24
geothermal energy, 21
incentives, financial, 55
market transformation, 110–13
privatization/restructuring, 28
transit impact development fee, 64
Campaign contributions and policy initiatives/options, 162
Canada, energy efficiency in, 22
Capacity building, 81–84, 91, 218
Carbon dioxide:
Annual Energy Outlook 2001, 137

Base/Clean Energy Scenarios, comparing U.S., 154–59
Brazil, 120, 183, 187
Clean Energy Scenario, Global, 227–30
Clean Energy Scenario, U.S., 147–48, 154–59
Earth Summit in 1992, 204
increasing levels, 8, 10
inequalities in production of, 15
IPCC Third Assessment Report, 17–18
Low Growth, Low Carbon Scenario, 16–17
market transformation, 95, 99
Netherlands, 107
overview, 6
regulations, 68
sequestration, 229–30
taxes, 61–62
United Kingdom, 126
United States, 132–35, 204–5
voluntary commitments, 65–66
Caspian Sea, 7
Cheney, Richard B., 162
Chicago and purchasing procedures, 75
China:
air pollution, 6
building energy codes, 68
capacity building, 82
efficiency, energy, 24
financing, 53, 54
inequalities, consumption, 14
information and training, 42, 70, 71
international policies/institutions, 197, 201
market transformation, 94–97, 108–10
regulations, 69
subsidies, 62
supply infrastructure, 34, 41
wind power, 41, 42
CHP, *see* Cogeneration systems
Cities for Climate Protection Campaign, 66
Clean Development Mechanism (CDM), 207
Clean Energy Scenario, Brazilian, 183–87
Clean Energy Scenario, Global, 225–32, 243–44
Clean Energy Scenario, U.S.:
appliance standards/building codes, 143–44
Base Scenario compared to, 148–60
carbon content standards for vehicle fuel, 147–48
coal issues/production, 146–47
cogeneration systems, 145–46
fuel economy, 139–40
overview, 138–39
Renewable Portfolio Standard, 142–43
research and development, 138–39, 145
system benefits trust fund, 140–41
tax incentives, 144–45
voluntary agreements, 141–42
Climate Technology Initiative, 193
Clinton, Bill, 50, 143
Coal issues/production:
as barrier to energy efficiency, 40
Base/Clean Energy Scenarios, comparing U.S., 150, 158
Clean Energy Scenario, Global, 244
Clean Energy Scenario, U.S., 146–47, 150, 158
depletion, resource, 14
global energy supply, 3
historical look at energy use, 2
market transformation, 97
Reference Scenario, 4
Cogeneration systems:
Brazil, 175–76, 179–81, 185, 187
Clean Energy Scenario, U.S., 145–46
efficiency, energy, 25
incentives, financial, 56
international policies/institutions, 200
Public Utility Regulatory Policy Act (PURA) of 1978, 63
regulations, 68
United Kingdom, 125–26
utilities as barrier to energy efficiency, 39–40
Colombia and population figures/issues, 237
Colorado and market reforms, 75
Combined heat and power (CHP), *see* Cogeneration systems
Communist nations, former, *see* Transition countries/economies; *individual countries*

Compact fluorescent lamps (CFLs):
  Brazil, 105
  California, 113
  future, towards a sustainable energy, 222, 223
  incentives, financial, 56
  purchasing procedures, 73
Competition and a more sustainable energy future, 218
Compressed natural gas (CNG), 62
Consumption, inequity and energy, 14–16, 38–39, 131–32, 165, 230–31
Contractors as barrier to energy efficiency, 36
Cookstoves, 70, 108–10
Corporate Average Fuel Economy (CAFE) standards, 100–101, 139–40
Costs:
  Base/Clean Energy Scenarios, comparing U.S., 152–54
  businesses as barrier to energy efficiency, 37
  efficiency, savings potential for energy, 23–24
  ethanol fuel, 120
  high energy, 5–6
  natural gas, 26
  oil issues/production, 134
  privatization/restructuring, 28
  research and development, 50, 51
  solar photovoltaic power, 20
  subsidies impacting, 63
  warming global, 8, 10
  weather events, global warming and extreme, 8
  wind power, 18, 20, 122
Czech Republic and capacity building, 83

Deaths caused by air pollution, 6, 7
Demand pull policies, research and development and, 51
Denmark:
  capacity building, 84
  incentives, financial, 55
  international policies/institutions, 197, 200
  market obligations, 79
  market transformation, 122–24
  research and development, 52, 122
  taxes, 61–63
  wind power, 43, 52, 122–24
Depletion, resource, 12–14
Developing countries:
  carbon dioxide, inequalities in production of, 15
  Clean Energy Scenario, Global, 230–31, 243–44
  Climate Technology Initiative, 193
  costs, high energy, 5
  efficiency, energy, 24–25
  electricity use, 15–16
  financing as barrier to energy efficiency, 37
  future, towards a sustainable energy, 224
  global energy supply, 3
  incentives, financial, 56, 57, 59–60
  inequalities, consumption, 14, 165
  information and training, 71
  international policies/institutions, 194, 195–99, 202
  market reforms, 76
  population figures/issues, 165, 237
  purchasing procedures, 73
  Reference Scenario, 4
  regulations, 67, 68
  research and development, 52
  solar photovoltaic power, 41
  supply infrastructure, 34, 41
  technology gap between rich and, 29
  urbanization, 29
  warming, global, 10
  see also Africa, sub-Saharan; Brazil; India; Transition countries/ economies; *individual countries*
Dominican Republic and financing, 53
Donor-driven programs as barriers to renewable energy use, 59–60
DuPont, 127

Earth Summit in 1992, 203–4
Ecological changes and global warming, 10–11
Economic context of energy:
  Base/Clean Energy Scenarios, comparing U.S., 154, 156–58
  Brazil, 120
  consumption, excessive, 38–39
  globalization, 27–28

technology gap between rich/poor nations, 29
see also Costs; Financing; Incentives, financial
Efficiency, energy:
Brazil, 171, 172–74, 177
cogeneration systems, 25
examples of technologies, 22–23
fuel cells, 25, 234–36
important worldwide resource, 21
international policies/institutions, 192, 210–13
Kyoto Protocol to the Climate Change Convention, 206
leapfrogging, 195
learning rates, 94
progress made in industrialized nations, 21–22
savings, cost-effective energy, 23–24
systems engineering approach, 23
transition countries/developing countries, 24–25
United States, 22, 132–35
see also Barrier listings; Clean Energy Scenario listings; Market transformation approach; Policy initiatives/options; Sustainable energy development
Electricity:
Annual Energy Outlook 2001, 136
Base/Clean Energy Scenarios, comparing U.S., 151
Brazil, 121, 170, 173, 185
developing countries, 15–16
European Union, 194
feed-in laws, 57
future, towards a sustainable energy, 224–25
incentives, financial, 55
inequalities, consumption, 14–16
market obligations, 77
market reforms, 76–77
market transformation, 102–6, 116–17
plans, energy, 86
politics as barrier to renewable energy use, 44
privatization/restructuring, 28
subsidies, 38, 43
Emissions cap schemes, 81
Employment and U.S. Base/Clean Energy Scenarios, 154, 156–58
Energy affecting all aspects of our lives, 1–2
see also Trends in energy use and their implications; *individual subject headings*
Energy Department, U.S. (DOE), 50, 98, 138, 145, 146
Energy Information Administration (EIA), 136–37, 149
Energy service companies (ESCOs), 53–54
Energy Star programs, 69, 70
Enron, 1, 162
Environmental Protection Agency (EPA), 145, 146
ESKOM, 77
Ethanol, 55, 117–22, 147, 178–79, 195
European Bank for Reconstruction and Development (EBRD), 54
European Union (EU):
air pollution, 6–7
capacity building, 82–83
fuel, vehicle, 132
future, towards a sustainable energy, 223
incentives, financial, 57–58
information and training, 69–70
renewable energy, 193, 194
research and development, 51
taxes, gasoline, 60–61
voluntary agreements, 64–65, 193
see also Germany; United Kingdom; *individual countries*
ExxonMobil Corporation, 162

Feed-in laws, 57
Financing:
as barrier to energy efficiency, 37–38
as barrier to renewable energy use, 42
Brazil, 170, 173–74
future, towards a sustainable energy, 217
market transformation, 115
policy initiatives/options, 52–54, 89
Finland and taxes, 61
Ford Motor Company, 139–40
Fossil fuels:
air pollution, 6
Bush, George H. W., 135
carbon dioxide, 8

Fossil fuels (*continued*):
Clean Energy Scenario, Global, 231
global energy supply, 3
radioactive contamination, 7
Reference Scenario, 4–5
subsidies, 38, 62–63
see also Coal issues/production; Natural gas; Oil issues/production
France and air pollution/public health, 7
Freight transport in Brazil, improving, 183
Fuel, vehicle:
Brazil, 177–78
Clean Energy Scenario, U.S., 147–48
economy standards, 40, 100–101, 139–40
ethanol, 55, 117–22, 147, 178–79, 195
future, towards a sustainable energy, 234–36
taxes, 60–62
usage, 132
Fuel cells, 25, 234–36
Fuelwood/firewood for energy needs, 3, 7, 8, 167
Future, towards a sustainable energy, 225–32, 243–44
geographic areas, specific, 223–25
lessons drawn from past experience, 216–20
overview, 215–16
population growth and lifestyle choice, 236–39
progress to date, 220–23
summary/review, 239–41
transportation challenge, 233–36

Gas, *see* Natural gas
Gas-fired power plants, Brazilian, 176–77
Gasoline, *see* Fuel, vehicle
General Motors Company, 140
Geological Survey, U.S., 12–13
Geothermal energy, 19, 21, 42
Germany:
incentives, financial, 57, 59
international policies/institutions, 200
plans, energy, 85, 87
politics as barrier to renewable energy use, 44
research and development, 50, 52
voluntary agreements, 65
wind power, 43, 44, 52
Ghana and market reforms, 76
Global Environment Facility (GEF), 54, 59, 77, 200–202, 210, 220
Globalization, 27–28
see also International policies/institutions
Global warming, 8–11
Government:
politics and obstacles to energy/renewable energy, 40, 43–44
relations between nations, energy affecting, 1–2
research and development funded by, 49–51
voluntary agreements with private sector, 64–66, 89–90, 141–42, 193, 217–18
voluntary commitments, 65–66
see also International policies/institutions; Policy initiatives/options; Regulations
Grameen Bank, 53
Grameen Shakti, 53
Greenhouse gases, 8
see also Carbon dioxide
Green Lights program, 69

Health, air pollution and public, 6–8
Historical look (1850-now) at energy use, 2–3
Honduras and financing, 53
Hungary:
financing, 54
international policies/institutions, 201
Hussein, Saddam, 12
Hydropower:
Africa, sub-Saharan, 21
Brazil, 170
Clean Energy Scenario, Global, 232
global energy supply, 3
plans, energy, 86–87
status of the technology, 19, 21

IBM, 127
Imports, Brazilian dependence on energy, 170–71
Incentives, financial:
as barrier to energy efficiency, 36
future, towards a sustainable energy, 217

market transformation, 100
policy initiatives/options, 55–60, 89
research and development, 51
India:
air pollution, 6
capacity building, 82
cookstoves, 109–10
costs, energy, 63
financing, 52
incentives, financial, 55, 60
indoor air pollution, 7
inequalities, consumption, 15
international policies/institutions, 201
market transformation, 109–10, 113–17
plans, energy, 85
privatization/restructuring, 28
subsidies as barrier to energy efficiency, 38
subsidies as barrier to renewable energy use, 42–43
Indonesia:
financing, 53
regulations, 69
Indoor air pollution, 7–8
Industrialized nations:
carbon dioxide, inequalities in production of, 15
costs, high energy, 5–6
efficiency, energy, 21–22
financing, 38, 42, 53–54
global energy supply, 3
historical look at energy use, 2
incentives, financial, 57
inequalities, consumption, 14
population figures/issues, 237–38
regulations, 67–68
research and development, 50, 52
security risks and dependence on oil, 11–12
technology gap between poor and, 29
see also California; Organization for Economic Cooperation and Development; United Kingdom; United States; *individual countries*
Inequity in energy consumption, 14–16, 38–39, 131–32, 165, 230–31
Infectious diseases and global warming, 10
Information and training:
as barrier to energy efficiency, 35–36
as barrier to renewable energy use, 41–42
future, towards a sustainable energy, 218
policy initiatives/options, 69–71, 90
technologies, information, 28–29
Innovation, *see* Technology/innovation
Integrated Model for Energy Planning (IMEP), 183
Integrated resource planning (IRP), 84–85
Intergovernmental Panel on Climate Change (IPCC), 17–18
International Council for Local Environment Initiatives, 66
International Energy Agency (IEA), 192–93, 209, 220
International Energy Efficiency and Renewable Energy Agency (IEEREA), 210–13
International Institute for Applied Systems Analysis (IIASA), 16–17, 226, 230, 243
International policies/institutions:
banks, multilateral development, 202–3
bilateral assistance/Global Environment Facility/United Nations, 200–202, 210, 220
cooperation among countries, 192–95
developing countries, 194, 195–99, 202
Earth Summit in 1992, 203–4
future, towards a sustainable energy, 220
Global Environment Facility, 54, 59, 77, 200–202, 210, 220
International Energy Agency, 192–93, 209, 220
International Energy Efficiency and Renewable Energy Agency, 210–13
Kyoto Protocol to the Climate Change Convention, 17, 61, 155, 187, 204–10
overview, 191–92
United Nations, 196, 202, 203–4, 209–10, 220
Investments in energy sector, Brazilian, 170, 173–74
see also Financing
IPCC, *see* Intergovernmental Panel on Climate Change
Italy, market obligations in, 79, 80

Japan:
capacity building, 82
efficiency, energy, 22

Japan (*continued*):
gasoline usage, 132
incentives, financial, 58–59
regulations, 66–67
research and development, 50–51
taxes, 62
Jeffords, Jim, 143, 147

Kazakhstan and air pollution, 7
Kenya:
information and training, 70
performance standards, 68–69
solar photovoltaic power, 41
Korea, South:
financing, 54
population figures/issues, 237
regulations, 67
Kyoto Protocol to the Climate Change Convention, 17, 61, 155, 187, 204–10

Lamps/ballasts, 51
see also Compact fluorescent lamps
Land use planning, 85
Lay, Ken, 162
Lead emissions, 6
Leapfrogging, 195
Learning rates, 94
Legislation:
Clean Air Act of 1970, 146–47
Northwest Power and Conservation Act of 1980, 86–87
Public Utility Regulatory Policy Act (PURA) of 1978, 63
Lieberman, Joe, 147
Lifestyle choice and a sustainable energy future, 236, 238–39
Loans, 53, 57
Long-term agreements (LTA) program, 106–8
Low Growth, Low Carbon Scenario, 16–17, 25

Market obligations, 77–81, 91, 217
Market reforms, 75–77
Market reserves, research and development and, 51
Market transformation approach:
Brazil, 102–6, 117–22
China, 94–97, 108–10
corporate examples, 127–28
Denmark, 122–24
experience curves, 93–94
future, towards a sustainable energy, 216
India, 109–10, 113–17
Netherlands, 106–8
overview, 48–49, 93–94
summary/review, 126, 128–29
United Kingdom, 124–26
United States, 97–102, 110–13
see also Policy initiatives/options
Methane, 8
Mexico:
financing, 53
incentives, financial, 56
international policies/institutions, 201
purchasing procedures, 73
regulations, 69
Microcredit, 53
Microelectronics revolution, 28
Minimum efficiency standards, 66–67
Mongolia and financing, Inner, 53
Multilateral development banks (MDBs), 202–3

National Academy of Sciences, U.S., 50, 238
Natural gas (NG):
Base/Clean Energy Scenarios, comparing U.S., 158
Brazil, 175–76
bridging fuel to help transition to renewable sources, 25–26
Clean Energy Scenario, Global, 244
costs, 26
depletion, resource, 14
historical look at energy use, 3
market transformation, 124–26
taxes, 62
United Kingdom, 124–26
Netherlands:
incentives, financial, 56
market transformation, 106–8
regulations as barrier to renewable energy use, 43
taxes, 61
voluntary agreements, 65
wind power, 43
Net metering, 63

Nevada, sustainable energy future in, 224
New Jersey:
future, towards a sustainable energy, 224
voluntary commitments, 65
New York Power Authority (NYPA), 72
New York state and purchasing procedures, 74–75
Nitrogen oxide, 81
Nitrous oxide, 6
Norway:
market reforms, 76
taxes, 61
Nuclear issues/power:
Bush, George H. W., 135
Clean Energy Scenario, Global, 244
global energy supply, 3
historical look at energy use, 3
research and development, 26–27
security risks, 12
stalled technology, 26
subsidies, 38
United Kingdom, 125

Ocean circulation and global warming, 10
Oil issues/production:
Annual Energy Outlook 2001, 136
as barrier to energy efficiency, 40
Base/Clean Energy Scenarios, comparing U.S., 150, 158
Brazil, 166
Clean Energy Scenario, Global, 230, 244
conflicts over oil, 2
crash in world oil price in 1986, 134
depletion, resource, 12–14
historical look at energy use, 2–3
security risks and industrialized nations dependence on oil, 11–12
subsidies, 38
Oregon and carbon dioxide regulations, 68
Organization for Economic Cooperation and Development (OECD):
carbon dioxide, 204
Clean Energy Scenario, Global, 230, 243–44
oil import dependence, 11
tax reforms, 61
Organization of Petroleum Exporting Countries (OPEC), 2, 136
Osama bin Laden, 12

Pacific Northwest and energy plans, 86–87
Particulate levels, 6, 7
Performance standards, 67–69
Persian Gulf region, 2, 136
Petrobras, 117
Petroleum production, *see* Oil issues/production
Philippines, building energy codes in, 68
Photovoltaics, *see* Solar photovoltaic power
Plans, energy, 84–88, 91
see also Future, towards a sustainable energy
Plutonium, 12
Poland:
capacity building, 83
incentives, financial, 56
Policy initiatives/options:
Brazil, 172–83
campaign contributions, 162
capacity building, 81–84
categories, 12, 47
financing, 52–54, 89
future, towards a sustainable energy, 216–20
incentives, financial, 55–60, 89
information dissemination and training, 69–71, 90
market obligations, 77–81
market reforms, 75–77
planning techniques, 84–88
purchasing procedures, 72–75, 90–91
regulations, 66–69
research and development, 49–52, 88–89
subsidies, 62–63, 89
summary/review, 88–91
taxes, 60–63
voluntary agreements, 64–66
see also International policies/institutions; Market transformation approach; United States
Politics and obstacles to energy efficiency/renewable energy use, 40, 43–44
Population figures/issues:
Brazil, 166
developing countries, 165, 237

Population figures/issues (*continued*):
future, towards a sustainable energy, 236–39
Prescriptive standards, 67
Pricing, *see* Subsidies; Taxes
Primary energy, xvi
Private sector:
future, towards a sustainable energy, 218
research and development, 51–52
restructuring/privatization, 28, 39, 76, 104–5
voluntary agreements with government, 64–66, 89–90, 141–42, 193, 217–18
Procurement, bulk, *see* Purchasing procedures
Public housing as barrier to energy efficiency, 36
Purchasing procedures:
as barrier to energy efficiency, 36–37
future, towards a sustainable energy, 218
policy initiatives/options, 72–75, 90–91
research and development, 51

al-Qaeda network, 12
Quality problems as barrier to energy efficiency/renewable energy, 34–35, 41

Radioactive contamination, 7
Reference Scenario, 4–5, 11, 25
Regulations:
automobiles, 69
as barrier to energy efficiency, 39
as barrier to renewable energy use, 43
building energy codes, 67–68
developing countries, 67, 68
future, towards a sustainable energy, 217
industrialized nations, 67–68
Japan, 66–67
market obligations, 77–81
market transformation, 95
minimum efficiency standards, 66–67
monitoring compliance, 90
research and development, 51
solar photovoltaic power, 68–69
updating standards, 90
Renewable energy:
Brazil, 171, 182
European Union, 193, 194
global energy supply, 3
historical look at energy use, 3
incentives, financial, 57–58
international policies/institutions, 192, 210–13
Kyoto Protocol to the Climate Change Convention, 206
learning rates, 94
sources, 18–21
status of major technologies, 18–20
see also Barrier *listings;* Biomass energy; Clean Energy Scenario *listings;* Hydropower; Market transformation approach; Policy initiatives/options; Solar photovoltaic power; Sustainable energy development; Wind power
Renewable Portfolio Standard (RPS), 77–79, 142–43
Rental property as barrier to energy efficiency, 36
Research and development (R&D):
appliances, 98
Brazil, 119
carbon dioxide, 229
China, 95
Clean Energy Scenario, U.S., 138–39, 145
cost reductions, 50, 51
Denmark, 52, 122
future, towards a sustainable energy, 217
government-funded, 49–51
international policies/institutions, 192
learning rates, 94
market reforms, 76
nuclear issues/power, 26–27
policy initiatives/options, 49–52, 88–89
private sector, 51–52
United States, 212
see also Technology/innovation
Royal Dutch/Shell, 127
Russia:
gasoline usage, 132
incentives as barrier to energy efficiency, 36
subsidies, 38, 62

Saudi Arabia, American soldiers in, 12
Sea level rise and global warming, 8, 10
Seattle (WA) and voluntary commitments, 66

Security risks and industrialized nations dependence on oil, 11–12
Shell Solar, 77
Singapore:
population figures/issues, 237
taxes, 63–64
Social context of energy:
Brazil, 120, 172
consumption, inequalities in energy, 14–16, 38–39, 131–32, 165, 230–31
international policies/institutions, 198–99
technology gap between rich/poor nations, 29
urbanization, 29
Solar photovoltaic (PV) power:
Africa, sub-Saharan, 41
Clean Energy Scenario, Global, 232
financing, 42, 52–53
future, towards a sustainable energy, 220–21, 225
incentives, financial, 58–59
India, 113–15
information and training, 71
international policies/institutions, 200
market transformation, 114
performance standards, 68–69
purchasing procedures, 74
quality problems as barrier to renewable energy use, 41
regulations, 68–69
research and development, 49–51
South Africa, 41
status of technology, 18–20
supply infrastructure, 41
Solar thermal power, 50
Soot, hazardous, 6
South Africa:
building energy codes, 68
indoor air pollution, 7
international policies/institutions, 194
market reforms, 77
quality problems as barrier to renewable energy use, 41
solar photovoltaic power, 41
Southern Company, 162
Spain, financial incentives in, 57
Sri Lanka:
financing, 53
plans, energy, 85
regulations, 69
Subsidies:
as barrier to energy efficiency, 38–39
as barrier to renewable energy use, 42–43, 59–60
Brazil, 119–20
fossil fuels, 38, 62–63
future, towards a sustainable energy, 217
policy initiatives/options, 62–63, 89
Sugarcane for ethanol, 118–21
Sulfur dioxide (SO_2), 6, 17, 81
Supply infrastructure as barrier to energy efficiency/renewable energy, 34, 40–41
Sustainable energy development:
IPCC Third Assessment Report, 17–18
Low Growth, Low Carbon Scenario, 16–17
natural gas, 25–26
nuclear issues/power, 26–27
opportunities, energy efficiency, 21–25
overview, 16
renewable energy sources, 18–21
summary/review, 29–30
see also Barrier *listings;* Clean Energy Scenario *listings;* Efficiency, energy; Future, towards a sustainable energy; Policy initiatives/options; Renewable energy
Sweden:
biomass energy, 20
international policies/institutions, 194, 200
plans, energy, 87–88
purchasing procedures, 72
regulations as barrier to renewable energy use, 43
research and development, 52
taxes, 61–62
wind power, 43
Switzerland, air pollution/public health in, 7
System benefits trust fund, 140–41
Systems engineering approach to energy efficiency, 23

Taxes:
automobiles, 63–64
as barrier to energy efficiency, 39

Taxes (*continued*):
as barrier to renewable energy use, 43
Clean Energy Scenario, U.S., 144–45
Climate Change Levy, 62, 125
fuel, vehicle, 60–62
incentives, financial, 57
policy initiatives/options, 60–63
privatization/restructuring, 76
technology/innovation, 55–56
transit impact development fee, 64
Technology/innovation:
Clean Energy Scenario, U.S., 144–45
Climate Technology Initiative, 193
future, towards a sustainable energy, 217, 222, 224–25
gap between rich and poor nations, 29
incentives, financial, 55–56
information technologies, 28–29
microelectronics revolution, 28
see also Research and development
Tennessee Valley Authority, 143
Terrorist attacks of September 11, 11–12
Testing and labeling programs, 69–70, 98
Thailand:
financing, 53
information and training, 71
international policies/institutions, 201
population figures/issues, 237
purchasing procedures, 73
quality problems as barrier to energy efficiency, 35
voluntary agreements, 64–66
Thermal power plants, Brazilian, 176–77
Toronto and voluntary commitments, 65–66
Toxic chemicals and burning fossil fuels, 6
Training, *see* Information and training
Trans-Alaska pipeline, 12
Transit impact development fee, 64
Transition countries/economies:
capacity building, 82–83
costs, high energy, 5
efficiency, energy, 24–25
natural gas, 25
Reference Scenario, 4–5
subsidies as barrier to energy efficiency, 38
see also Developing countries; *individual countries*
Transport planning, 85, 87–88, 233–36
see also Automobiles; Fuel, vehicle
Trends in energy use and their implications:
air pollution, local/regional, 6–8
depletion, resource, 12–14
economic/social context of energy, 27–29
1850, energy use since, 2–3
global energy supply, 3
inequity in consumption, 14–16
Reference Scenario, 4
security risks, 11–12
summary/review, 29–30
warming, global, 8–11
see also individual subject headings

Ukraine:
capacity building, 82–83
costs, energy, 5, 63
United Kingdom:
efficiency, energy, 22
future, towards a sustainable energy, 223
gasoline taxes, 60–61
incentives, financial, 56
market obligations, 79–80
market reforms, 76
market transformation, 124–26
quality problems as barrier to energy efficiency, 35
taxes, 62
United Nations, 196, 202, 203–4, 209–10, 220
United States:
Base/Clean Energy Scenarios, comparing, 148–60
business-as-usual base scenario, 135–37
efficiency, energy, 22, 132–35
future, towards a sustainable energy, 223
incentives, financial, 55
inequalities, consumption, 14, 131–32
international policies/institutions, 196, 209–10
Kyoto Protocol to the Climate Change Convention, 204–5, 208–9
market transformation approach, 97–102, 110–13
politics as barrier to renewable energy use, 44

purchasing as barrier to energy efficiency, 37
Renewable Portfolio Standard, 78
research and development, 50–52, 212
summary/review, 161, 163
sustainable energy development, 137–48
voluntary agreements, 65
see also California; Industrialized nations
United Technologies Corporation, 127–28
Units of energy, xv–xvi
Uranium, 7, 12
Urbanization, 29
Utilities:
as barrier to energy efficiency, 39–40
as barrier to renewable energy use, 43
Brazil, 121
incentives, financial, 55
market obligations, 77
privatization/restructuring, 28
see also Electricity; Oil issues/production

Vietnam:
financing, 53
international policies/institutions, 201
Volatile organic compounds, 6
Voluntary agreements between government/private sector, 64–66, 89–90, 141–42, 193, 217–18
Voluntary commitments and governmental authorities, 65–66

Warming, global, 8–11
see also Carbon dioxide
Waxman, Henry, 147
Weather events, global warming and extreme, 8, 10
Wind power:
Brazil, 181–82
capacity building, 84
China, 41, 42
Clean Energy Scenario, Global, 232
Denmark, 43, 52, 122–24
financing, 53
future, towards a sustainable energy, 220, 225
Germany, 43, 44, 52
global energy supply, 3
incentives, financial, 55, 57
information/training as barrier to renewable energy use, 42
international policies/institutions, 197, 200
market obligations, 78
market reforms, 75–76
market transformation, 115, 122–24
politics as barrier to renewable energy use, 44
regulations as barrier to renewable energy use, 43
research and development, 49, 50, 52
status of the technology, 18–20
supply infrastructure as barrier to renewable energy use, 41
Women:
consumption inequalities, 16
international policies/institutions, 198–99
World Bank:
as barrier to renewable energy use, 44
capacity building, 84
future, towards a sustainable energy, 220
incentives, financial, 59
international policies/institutions, 203
market transformation, 105
World Energy Council (WEC), 16, 226, 243
World Energy Outlook 2000, 4, 11
World Health Organization (WHO), 6, 7
World War II, 2–3

Zimbabwe:
international policies/institutions, 200
solar photovoltaic power, 41
subsidies, 60

Island Press Board of Directors